Nursery Management:
A Complete Guide for Nurseryman

Nursery Management:
A Complete Guide for Nurseryman

Manas Ranjan Nayak
Scientist (Forestry)
Krishi Vigyan Kendra, Koraput-763002
OUAT, Bhubaneshwar
Odisha

SSPH

SATISH SERIAL PUBLISHING HOUSE
403, Express Tower, Commercial Complex
Azadpur, Delhi-110033 (India)
Phone : 011-27672852, Fax : 91-11-27672046
e-mail : info@satishserial.com, hkjain1975@yahoo.com
Website : www.satishserial.com

Published by :

SATISH SERIAL PUBLISHING HOUSE
403, Express Tower, Commercial Complex, Azadpur, Delhi-110033 (INDIA)
Phone : 011-27672852 Fax : 91-11-27672046
E-mail : info@satishserial.com, hkjain1975@yahoo.com

ISBN 978-93-86200-67-9

Composed, Designed & Printed in India

Dr. Niranjan Senapati
Principal Scientist, Centre for Pulses Research, Berhampur
Orissa University of Agriculture and Technology, Bhubaneswar

FOREWORD

Demand for seeds and seedlings of silviculture, horticulture and vegetable species has been is increasing over the years. The special plantation drives and greening activities have substantially increased the demand for flowers, plantation crops, medicinal and aromatic plants seedlings. Ease of availability of quality planting material (QPM) at reasonable costs, however, is a challenge. At present only up to 30-40% demand for planting material is being met by the registered nurseries; the rest are met from the unorganized sectors, implying the need for establishing more nurseries in the organized sector.

Now a days the environment and climate change issue are of great concern and for mitigating the adverse effects, we have to propagate the healthy seedling of different tree species. The healthy planting material comes from nursery and hence we should take care of seedling in the nursery stage. This publication covers the principles and practices in the production of quality and superior planting materials from seeds to seedlings grown in the nurseries. The aim of this manual is to familiarize the nursery operators, farmers and institutional nurseries with, and to get them involved in, the project techniques of propagation and establishment of nurseries.

This book entitled **"Nursery Management: A Complete Guide for Nurseryman"** will serve as a ready reference and help entrepreneur, farmers, student nursery industries as well as scientists to produce healthy seedling. I appreciate of Dr. M. R. Nayak, Scientist (Forestry) for their efforts in compilation of this book.

Bhubaneswar
11th Nov. 2017

N. Senapati
Principal Scientist

Preface

The aim of this book is to provide information and guidance on how to start a nursery and the steps involved in growing quality tree seedlings. The book is therefore of special interest not only to nursery workers and nursery supervisors but also to students, scientists, foresters, horticulturists and those involved in planning and managing the production and use of seedlings for any of the above mentioned roles and benefits.

The main objective of this book is to help the readers, especially B.Sc. and Diploma students of agricultural graduates, to quickly grasp the facts comprehensively and systematically from various subjects of Nursery like Useful tools and Equipments, Choosing the nursery site, Design and layout, Nursery operation, Seedling production, Plant protection in Nursery, Growing media, Plant propagation structure, Nursery management tips, Nursery law and regulation etc. This book consists of 16 chapters on various aspects of Nursery. This book will meet out the growing need of students to acquire the up to date information in the various subjects of Nursery for getting success in Nursery.

We hope that the information provided and compiled in this book would be useful for students, nursery man and it serves our purpose. Readers are welcome to point out errors and omissions, if any, in the book and also requested to send their valuable suggestions for further improvement of quality of this book.

M.R. Nayak

About the Author

Dr. Manas Ranjan Nayak is a Scientist (Forestry) working in the Krishi Vigyan Kendra, OUAT, Bhubaneswar, Odisha (India). He has more than 2 year of experience in research and extension in the field of forestry and agroforestry. He received his B.Sc. (Forestry) and M.Sc. (Forestry) from Orissa University of Agricultural and Technology, Bhubaneswar, India and Ph.D Forestry (Agroforestry) at Navsari Agricultureal University, Gujarat. He was the recipient of Merit Scholarship during M.Sc., University Gold Medal from OUAT, Bhubaneswar, INSPIRE Fellowship, from DST, Govt. of India. He has qualified ICAR-SRF, ICAR-NET examination in agroforestry. He has more than 20 research papers published in various national and international journals and also published popular articles on forestry, Agroforestry and environmental aspects. He was participated in several national and international conferences, seminars, workshop/ training, summer school and presented oral and poster presentation in the thematic area of forestry, Agroforestry, Biodiversity and climate change. He wrote 2 books: "Agrisilviculture in Odisha-Food Crops and Soil Fertility" (ISBN: 978-3-659-58141-0) and "Forestry Solution" (ISBN: 818399098-3)

Acknowledgements

I am deeply indebted to all colleagues of the Krishi Vigyan Kendra and Regional Research Technology Transfer Station, Semilguda, Orissa University of Agriculture and Technology, Bhubaneswar (India) who inspired us to crystallize our experience in the form of the present book.

I am specially thankful to my parents, my wife (Rakhi), My son (Sibansh) for giving constant support for writing this book.

Our deep sense of gratitude and sincere indebtedness to Honourable Vice-Chancellor, OUAT, Bhubaneswar, the Dean, Extension Education, OUAT, Bhubaneswar, Dr. N. Senapati, Principal Scientist, CPR Berhampur, Dr. D.B. Jadeja, Retd. Professor and Head, College of Forestry, NAU, Gujarat, Mr L. Nayak, Jr. Agronomist, RRTTS, Semiliguda, Smt J.R. Maharana, Senior Scientist and Head, KVK, Koraput, Dr. A.K. Patra, Associate Professor, IFS, OUAT, Bhubaneswar who are the source of inspiration, which enable me to write this book.

Special thanks to is extended to my loving friends Lingraj, Sunita Madam, Biman Sir, Laxmi Bhai, Prasanta Bhai, George, Bibhuti Sir for rendering necessary help for the writing the manuscript as and when required during publication of this book.

Finally, Heartful thanks goes to our family for their active co-operation and understanding, which enabled us to devote so many long hours to what, at times, seemed an endless task.

M.R. Nayak

Contents

Foreword .... *v*
Preface .... *vii*
Acknowledgement .... *ix*

1. Introduction 1
2. Useful Tools and Equipments 7
3. Choosing Nursery Site 21
4. Design and Layout 31
5. Nursery Operations 41
6. Producing Plants from Seed 51
7. Producing Plant Vegetatively 79
8. Plant Protection in Nursery 109
9. Plant Propagation Structures 131
10. Growing Media 145
11. Nursery Record 157
12. Nursery Business 165
13. Nursery Law and Regulation 177
14. Propagation Techniques for Important Tree Species 193
15. Nursery Management Tips 199
16. Glossary 205

1

Introduction

Trees/Plants are essential to our lives and are an integral part of our landscape. They play many roles and confer a multitude of benefits to society such as food, shelter, shade, timber, natural beauty to the countryside, conservation of our soil and water resources and improved air quality. With the recent approval and launch of a National Forest Management and Conservation Plan, and the greater awareness of forest and tree values, indications are that there will be an increased demand for tree seedlings to ensure that these benefits continue to contribute to the nation's well-being.

At present the need to plant trees on farms is on the increase. It is difficult, however, for smallholders to access at the right time, in the right quantities and of high quality - the trees that they want to plant. In order to meet present and future demand for planting material, there is a need to promote on-farm and community tree nurseries. Such nurseries can be owned and managed by individual farmers, by self-help groups, by schools and/or by a range of other local institutions. They provide income-generating opportunities, act as models for further nursery development, provide seedlings more cheaply to planters, and can raise the particular species that local people are interested in.

A **Nursery** is a place where plants are propagated and grown to plantable size. To make a good plantation, good nursery stock is essential. Major mortality of seedlings in their plantations is due to the wrong size or poor health of the seedlings at the time of planting. In addition, poor seedlings are likely to have slower growth, to be less able to compete with weeds, and to be more liable to damage by insects and pests. Further, in a poor nursery, fewer seedlings will be raised from a given quantity of seed, and there will be considerable waste of money and time. After planting, the plants are immediately exposed to a harsh environment, and are at their most susceptible to damage from drought, grazing, fire, insects etc. Thus sound nursery practice is the foundation of a successful plantation scheme.

By definition, a nursery is an enclosed or fenced area designated for the short term growing of planting materials in anticipation of moving them to other sites or locations at their appropriate maturity. The plants which are usually grown in the nursery may be seedlings (sexually propagated) or asexually propagated planting materials of agricultural, horticultural and forest trees or any special plant species required for small and large scale planting at a new location. A nursery is a place where plants are propagated and grown to usable size for sale to others or self use.

A Good Nursery has three important aspects to look into: a) the Nursery Infrastructure; b) Production System and Quality Parameter of Planting Material Adopted; and c) Adoption of Good Nursery Management Practices. With these three aspects put together, a Model Nursery can be defined in terms of Nursery Infrastructure, Production System and Quality Parameter of Planting Material and Good Nursery Management Practices in a Comprehensive Manner.

Seedling production is one of the key steps in scaling up or domestication of any species. Each step has to be properly planned and implemented. The way seedlings are handled and managed in a nursery contributes to their survival rate after planting and their subsequent growth performance. Improving seedling quality correlates positively to their survival, growth and productivity. Seedling quality is governed by the genetic make-up of the parent trees and the physical growth of the seedlings. Several types of nurseries exist: individual or private, community or group, central or research, commercial and training nursery.

Nursery practices must be consistent and the various techniques closely integrated. If one element in the chain is lacking there will be a negative impact on seedling quality. Good quality seedlings cannot be produced without care and tending. Nursery plants need to be protected from extremes of environmental conditions until they are strong enough to withstand them. To ensure high quality of seedlings and to provide more opportunities (income, technology transfer), local people are encouraged to establish small-scale community nurseries.

What is a seed?

A seed is a small embryonic plant enclosed in a covering called the seed coat, usually with some stored food. It is the product of the ripened ovule of gymnosperm and angiosperm plants, which occurs after fertilization and some growth within the mother plant. The formation of the seed completes the process of reproduction in seed plants (started with the development of flowers and pollination), with the embryo developed from the zygote and the seed coat from the integuments of the ovule. All seeds have different sizes, shapes and colours.

Seeds of woody plants exhibit a great range of variation in shape, size, colour and behaviour. The most essential factor for the success of plantation is the ready availability of quality seeds. The quality of seed is responsible for the future performance of each and every seedling. Poor quality seeds may have the following problems:

- Low germination percentage
- Poor emergence
- Poor survival
- Poor adaptability to the site
- Susceptibility to disease and pests
- Poor growth
- Low productivity

Characteristics of good seeds

- Must be well-ripened, healthy and true to type.
- Must be pure and free from inert materials and weed seeds.
- Must be viable and have good germination capacity.
- Must be uniform in texture, structure and look.
- Must not be damaged, broken or affected by pests and diseases.

What is a seedling?

A seedling is a young plant that is grown from a seed. It is any young plant, especially one grown in a nursery for transplanting. Seedling development starts with germination of the seed.

What is plant propagation?

Plant propagation is the process of creating new plants from a variety of sources: seeds, cuttings, bulbs and other plant parts. Plant propagation can also refer to the artificial or natural dispersal of plants. The simplest method of propagating a tree vegetatively (non-sexually) is rooting or taking cuttings. The most common method of propagating fruit trees, suitable for nearly all species, is grafting or budding a desired variety onto suitable rootstocks.

Importance of Nursery and its Role

The availability of quality planting materials is the foundation of crop production. If the plant materials are genuine, reliable and have guaranteed performance, they will certainly give higher yield and top-quality crop. Therefore, nursery is a place where from one can get plant materials of true-to-type. Now, nursery is undoubtedly a very paying business.

The importance of a nursery is mentioned below

- Seedlings and grafts are produced in nursery from which the fruit orchards, ornamental gardens and plantation can be established with minimum care, cost and maintenance.
- The nursery planting materials are available at the beginning of the planting season. This saves the time, money and efforts of the farmers to raise seedlings.
- There is a wide scope for fruit orchards, ornamental, vegetable, and landscape gardens at public places, highways and co-operative housing societies.
- It assures the production of genetically improved quality planting material.
- It provides employment opportunities for technical, skilled, semi-skilled, unskilled labour.
- They are an important source supplying the seedlings for meeting the fruit, pulp and paper, fuel wood, timber and other demands of the industries.
- Some species are not annual good seed bearer, but need to be planted annually. To meet the need of seedlings of such species, nursery is important.

- Slow growing species need a nursery to be planting out to avoid competition.
- Roadside and urban plantation always need a nursery for their plantation.
- The best method of introduction of exotics is only by nursery.
- Planting of nursery-grown plants is the surest method of artificially regenerating poor and barren sites.
- Casualty replacement is only possible by the plants grown in the nursery.

Seed propagation is the principal mode of plant production in temperate and tropical silviculture. In managing nursery operations based on seed propagation, the main objective is to obtain good germination and provide optimum conditions for their survival and growth into strong healthy trees. During germination and the early establishment phase, seedlings are susceptible to stress from improper watering, mechanical damage and infection and advice is provided on how to improve conditions so that seedlings can survive and pass through this vulnerable stage as efficiently as possible. The often overlooked need to gradually apply stress to harden seedlings in preparation for the harsher field environment is also covered.

During the last three decades, vegetative reproduction using parts of growing plants such as branch and tip cuttings, roots and leaf portions has become increasingly popular because of the difficulty of some plants to produce viable seed. This popularity has been enhanced by the development and perfection of mass production techniques involving plant cells (tissue culture) resulting in the world-wide availability of improved varieties of ornamental and horticultural seedlings. Because of its importance in producing seedlings for agroforestry and horticulture, vegetative reproduction is covered in Chapter 7.

To get success in nursery one has to maintain the standard of the quality of produce. The demand of the nursery products largely depends on their quality. So a good nursery should have the following qualities:

- A nursery should supply only the genuine plant materials in regard to the variety and species of the crop plant.

- Plant materials produced should be healthy, true-to-type and free from any type of disease and insect attack.
- The nursery should have got registration from the proper registration authority.
- The cost of the nursery products should be comparatively cheaper and acceptable to the customers.
- The nursery should be prompt in regard to supply of the plants and other correspondences.
- Plants that are sold from the nursery must be properly packed, so that it can stand the transport shock well and reach to the customer in fresh condition.
- Each plant should be properly labelled before handling over to the customers.
- Nursery man should be honest, sincere and well-behaved.

Many factors influence the way forests are managed in different parts of the world and the way forest nurseries fit into local forestry operations. The scale and character of forest operations and the demand for the supply of planting stock are interdependent. Owning a nursery/nurseries helps in matching demand with production of planting material and controlling its quality. It creates opportunities for employment and builds local expertise in operating forest nurseries. On the other hand, long-term flexibility is reduced by the weights of investments and commitments made during nursery development. Typically, large forest companies and government managing forests tend to establish their own nurseries, while small scale forestry operators tend to purchase the planting stock. Many independently owned nurseries have specialized in production of planting stock for sale only.

2

Useful Tools and Equipments

Introduction

Nursery managers realize that maintaining trees as a renewable resource requires highly productive nurseries. Modern tools and equipments can offer an efficient means of increasing nursery productivity without sacrificing seedling quality or employee safety. But staying abreast of developments in nursery equipment can be difficult. The equipment used by staff in a nursery can make a great difference to overall productivity. A lack of spending in this area can make it impossible for staff to do an effective job, even if they are very skilled and competent staff. On the other hand, excessive spending in this area can result in a significant waste of money that could be better spent elsewhere.

For carrying out day-to-day routine cultural operations in the nursery, various tools, implements and accessories are required. Some tools are simple and are used for simple operations, whereas for carrying out specific operations, special types of equipments are required.

This chapter informs nursery managers about current developments in nursery equipment and offers ideas about how custom built and commercial nursery tools and equipments can serve common functions.

Good management of tools and equipment involves three things:

1. Choosing the right type and quantity tools and equipment in the first place.
2. Making sure tools and equipment are used and maintained properly.
3. Upgrading or replacing tools and equipment at the right time.

Design of Tools

When a piece of equipment is made, it is usually made for a particular job. Nursery tools are usually used frequently, maybe every day, whereas tool for the home garden are only used occasionally. You should choose tools and equipment which are strong enough to withstand constant and sometimes heavy use.

- Strong tools will take greater physical stress and handle heavier jobs without breaking. If you buy a cheap tool built with low quality materials it might not last till the end of the first day, particularly if you are doing heavy work.
- Sharp tools put less stress on the tool, and less strain on the user.
- Long handles give greater leverage and increased reach, putting less strain on the user's body. Long handles, however, are sometimes impracticable, especially where the tool will be used in confined spaces.
- Some tools are more expensive because long hours of thought have gone into their design, not just because they use better materials. Tools which do the job better and more easily may be more expensive ones, but they are worth it.
- Heavy clay or rocky soils are likely to put more strain on tools and equipment such as spades and cultivators - you will need better quality and more heavy duty tools.
- Metal tools made with stainless steel or aluminium do not corrode like those made with other metals.

Deciding what is good to buy: Before you buy a tool or piece of equipment you need to decide 'do you really need it?' Too often equipment is bought simply on a whim, or because every one else has one or because it seems a good idea at the time. Deciding to buy should be based on a genuine need for that piece of equipment, for such reasons as significantly reducing workloads, or the time involved in tasks, allowing you to expand the types of activities you undertake in the nursery, for safety reasons, or for improving the quality of your work. The need for that equipment then needs to be balanced against the cost, both initial and ongoing, and the operational requirements of that equipment. Is it really a worthwhile proposition? This must be decided on such factors as:

- Does it do the job that you require of it?
- Initial cost - how much does it cost to buy?
- Ongoing costs - this includes maintenance, parts, fuel, & insurance.
- Reliability - does it break down a lot?
- Longevity - how long will it last?
- Safety to use.
- Availability of parts and servicing.

Protective Tools for Nursery Work

Footwear: Apart from saving your casual shoes or trainers from getting dirty or damaged, having a pair of heavy duty shoes or boots will also protect your feet from stones, falling items or tools, so it's worth investing in a sturdy pair of gardening shoes or wellington boots.

Nursery Aprons: Aprons not only protect your clothes from dirt but many are also highly functional with additional pockets and pouches which can hold things from seed packets to hand tools - plant labels to nursery twine. They also come in handy for your mobile phone or keys.

Gloves: A pair of gardening gloves is a must to keep your hands protected from cuts and abrasions. There are many varieties available from light-weight cotton gloves to thick waterproof heavy duty gloves so it may be worth investing in a pair of each type.

Hand Tools

Spade: The basic gardening tool is the spade, used for digging, planting, hoeing, shovelling, and cutting lawn edges. Long-handled spades have more leverage when digging, but the short-handled spade encourages the user to grip the handle lower down more under the load, and to use the leg muscles more.

A long handled tool traditionally used for digging, shovelling soil and compost. They are available in various sizes and can often be bought as a set together with a garden fork.

Rake: A rake can be used to make a fine seedbed, to open and close seed-drills, to remove lawn clippings and leaves, to tear out moss and dead grass, and to freshen up gravelled areas and flower beds. Long, even strokes of a rake are best, so a long handle is essential. The head should not be very wide, or too narrow. The teeth should be straight, or only slightly curved, and set not too far apart. Do not confuse with a spring tine rake which has little use when growing crops.

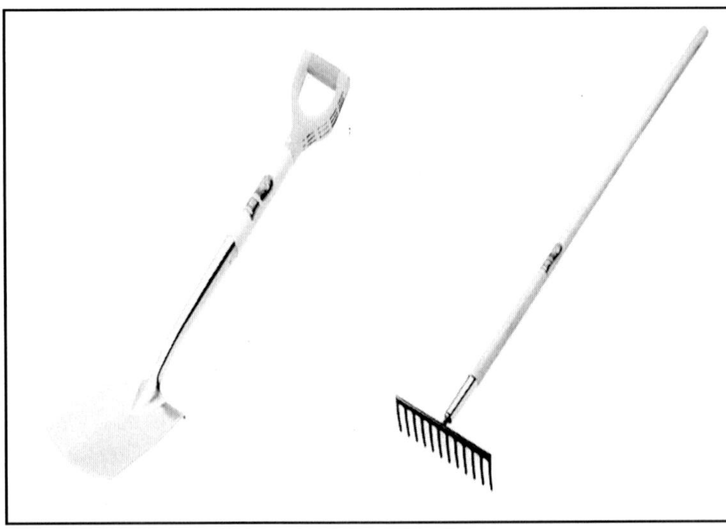

SPADE RAKE

Hoe: There are basically two types of hoe: push hoe and draw hoe. With a push hoe, the user moves backwards on to the un-hoed ground, and thus avoids walking on the newly-hoed weeds. With a draw hoe, the user moves forward towards the un-hoed ground and walks on the newly-hoed area. Raking off the hoed weeds is essential to prevent re-rooting.

A draw hoe is easier to use because the weight of the arms is brought down to help with the weed cuting action, whereas the pushing action of a push hoe requires the weight of the arms to be lifted at the same time.

A draw hoe can also be used in a standing position with a sweeping brush action to control light weed growth, and this gives the back a rest. A good hoe of either type should be of solid material and have a long handle. It is the main weapon against weeds.

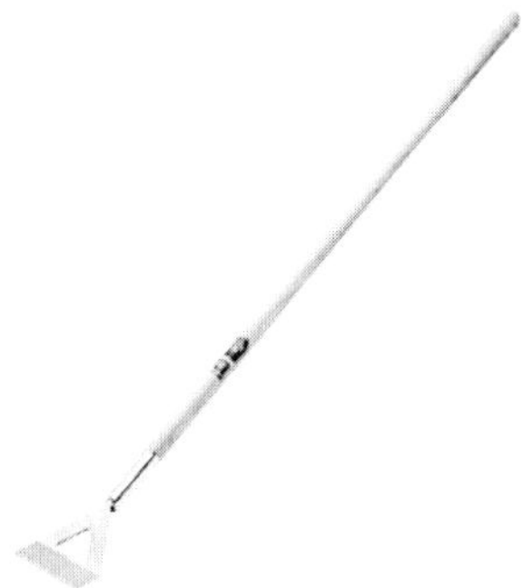

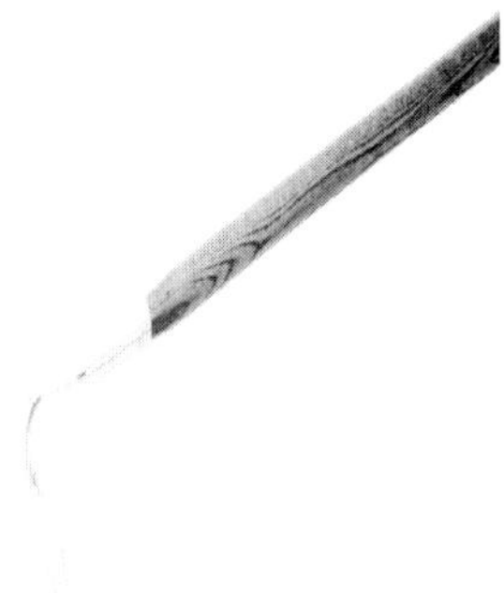

PUSH HOE DRAW HOE

Trowel: A planting trowel is essential for planting bedding and vegetable plants. The handle should be fairly short, and broad and smooth at the end for comfort in the palm of the hand. The blade should come up close to the end of the handle.

Ideally, the blade should be slightly curved – making it easier to take out a planting hole quickly, but not so curved as to have moist soil stick to it. The trowel can also be used to assist hand-weeding. Little hand-forks are useful for this.

Secateurs: The secateurs, or pruners, are the basic pruning tool. Essential for roses or fruit trees, it will be needed for shrubs on occasion, too. Secateurs are ideal for dead-heading, and for cutting flowers and foliage for indoor use.

The type with a curved blade cause less damage to the bark of the pruned stem. Buy good quality secateurs because, being better designed, they are easier to use, and last longer. A comfortable hand-grip is important.

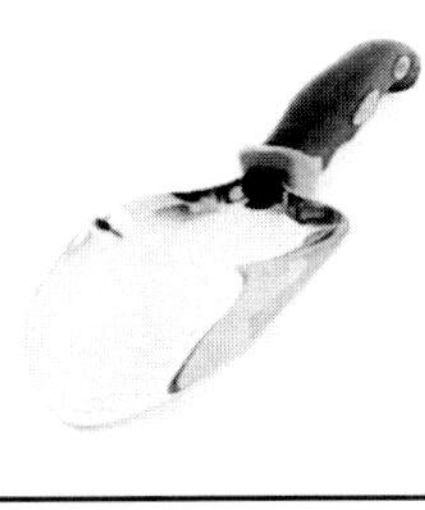

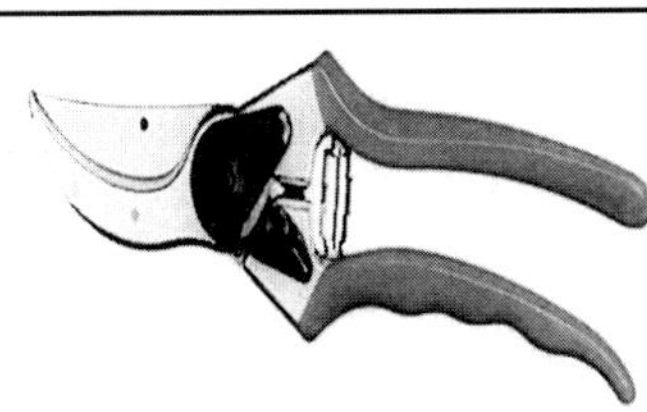

TROWEL SECATEURS

Hedge-clippers: A hedge clippers is essential if there is a hedge to be trimmed, but they can also be used to keep lawn edges neat. Hedge-clippers should not be used on shrubs, except in certain cases, such as heather, broom and lavender where there are a lot of shoots close together.

It is useful for trimming rockery perennials after flowering. A big range of sizes and designs is available. Choose one with a lock-nut assembly, which allows it to be tightened properly.

Watering can: A watering can has a vital role in ensuring the establishment of young plants of all kinds. It can also be used to apply weed killers on paths and lawns. It can be used as a substitute for a sprayer to apply insecticides too. Most watering cans are sold in a 10 litre size, which is ideal not too heavy when full.

Use a watering can rose of a fine droplet size for a wider range of uses, including watering seedlings and applying weed killers. If the can is used frequently for weed killing, it might be advisable to have a separate can for that purpose to avoid mishaps if it not properly washed out.

HEDGE-CLIPPERS

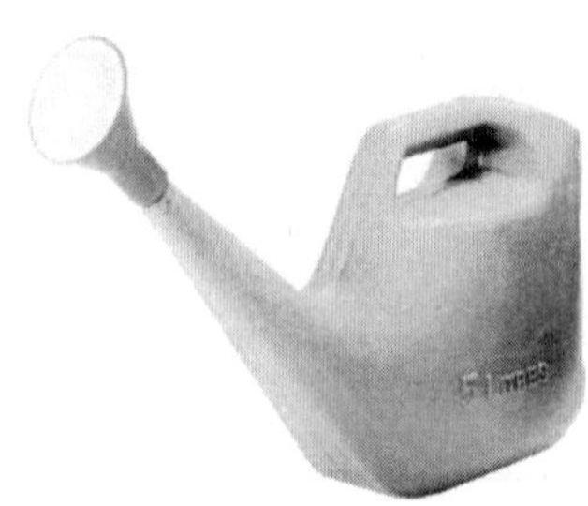

WATERING CAN

Garden fork: A garden fork can be used for digging; it is an essential piece of equipment on stony ground. It is useful for picking up debris such as prunings. For compost-making, it is a tool without which it is difficult to keep the heap tidy. Though not suitable for digging, a dung fork is more useful for the other tasks mentioned.

Lopping shears: A long-handled lopping shears is useful if a lot of roses, or fruit bushes are grown, and it is handy for pruning shrubs too, making it much easier to reach down into the tangle of branches and easier to reach high branches too. It will deal with branches too large for the secateurs, because the long handles give the user more leverage.

Edging shears: An edging shears is very useful if there is a lot of lawn edging to be cut around flower beds and pathways. The long handles make this job easier on the back. Awkward to use at first, the edging shears takes a little practice.

Riddle or garden sieve: A riddle is necessary if garden soil must be sieved for home-made composts, it is very useful for sieving out coarse pieces of moss peat.

Garden line: A simple piece of equipment for getting lines of vegetables, bedding, roses and hedging straight. A good line can be made from coloured nylon builder's line, tied to and wound around two pieces of metal rod or hardwood, about 30 centimetres long.

Pruning saw: A pruning saw is a very useful tool in an established garden with trees and shrubs, it is essential if sizeable branches have to be removed. A bow-saw is necessary for tree branches and can be used to cut up firewood as well.

Manual Potting Equipment

Smaller nurseries may find that their level of production would not justify the purchase of specialised potting equipment. The minimum potting equipment required could include:

- Solid washable workbenches (i.e. galvanised metal top) suited to propagation and potting activities with storage underneath for pots and trays and shelves above for labels, markers and propagating equipment.
- An adjacent under cover media storage area with a concrete base this should be located so it is accessible from both sides; for media delivery, and to enable the media to be shovelled onto potting benches.
- A mobile bench/potting cart is also a cheap investment. This allows the potting media to be moved to where it is needed and also seconds as a potting bench. The cart can also be used to move potted tubes to the growing-on area.
- A rotating table (lazy Susan type) suited to potting up plants.
- Nursery trolleys for moving plants.

GARDEN FORK

LOPPING SHEARS

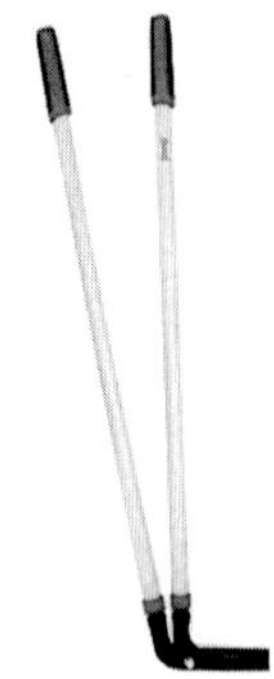

EDGING SHEARS

GARDEN SIEVE

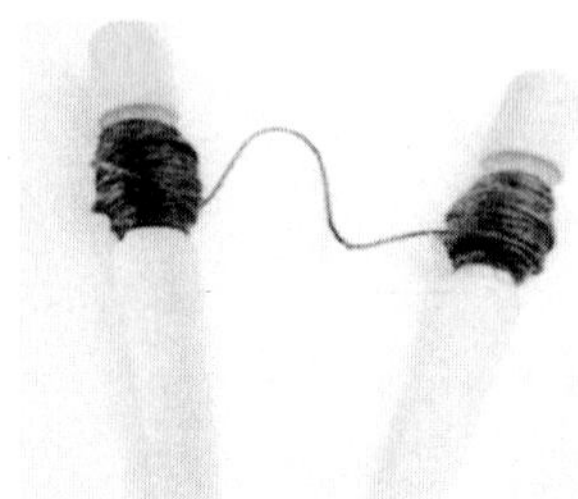

GARDEN LINE

PRUNING SAW

Nurseries with medium production runs may consider using:

- A potting station or potting hopper and bench. The potting station usually incorporates a gravity feed hopper that can be bulk or conveyor filled it is also easily adjustable to suit various pot sizes. This type of station has no moveable parts making it easy to maintain and reliable to use. The hopper and bench type is usually made from heavy galvanised steel. The hopper is bulk filled and tubes trays and pots of all sizes can be straight filled. The potting station usually incorporates a gravity feed hopper that can be bulk or conveyor filled it is also easily adjustable to suit various pot sizes. This type of station has no moveable parts making it easy to maintain and reliable to use. The hopper and bench type is usually made from heavy galvanised steel. The hopper is bulk filled and tubes trays and pots of all sizes can be straight filled.
- Storage hoppers
- Trailers that are specifically built to suit narrow and confined spaces.

- Small tractor particularly useful in the medium to larger type operation where a lot of material, equipment and stock is constantly being moved.
- Nursery trolleys are an essential item in every nursery. These are three or four-wheeled trolleys with a wide, shallow tray which is primarily used for moving plants around the nursery (most trolleys will hold both pots and seedling trays). The trolley tray is covered with a metal mesh to allow water to freely drain away when plants in the trolley are watered before unloading. Several trolleys should be conveniently located in both production areas for nursery workers and sales areas for customers to transport their selected plants in.

Hand Trolleys. Upright hand trolleys are useful for moving many things about the nursery including boxes of materials (eg. chemicals, pots, stationary, tools, etc.) when delivered and large planter tubs when a forklift or tractor isn't available. They are particularly useful in retail nurseries or at shows and exhibitions where a small upright trolley can sometimes be more manoeuvrable than a larger nursery trolley in a confined space.

Additional Potting Equipment for Larger Nurseries

Soil mixing equipment: Cement mixers (small or large) are sometimes used for mixing potting media in those nurseries which make their own mixes. In some larger nurseries large amounts of soils, composts, mulches or growing media may be mixed and/or transported using a front-end loader. Ideally these materials should only be mixed and stored on a clean surface such as concrete so that they are not contaminated by soil particles and soil-borne pest and diseases from underlying soil.

Conveyor Belts: Conveyor belts can be used to move plants away from work areas to growing on areas as well as moving media from hoppers to potting machines or benches. They are not viable in small nurseries, but some large nurseries have found that they can contribute towards improved productivity.

Potting Machines: Potting plants is extremely labour intensive and many nurseries have found that automated potting machines have significantly improved nursery efficiency. There are a number of different models available, including machines which do the following:

a) Controllable volume system - the amount of soil which is poured into the pot is controlled by pushing a foot pedal/hand button. Soil will continue to be delivered until the foot pedal/button is released.

b) Pre-set volume - the volume of soil is pre-set, although this can be changed for different sized pots, plants, etc.

c) Combination of the above - more sophisticated machines can be set to either method.

Pots can be moved along a conveyor belt or set on rotating benches. The soil may be fed through a central hopper or along a conveyor belt.

Seeding machines: There are a range of different types of seeding machines used by nurseries. Most feed seeds into soil or potting media by a vacuum mechanism (eg. seeds are picked up when a vacuum is created, then dropped in a desired position when the vacuum is disengaged). Plants grown in the field may be direct sown into nursery rows.

Plants grown in containers can be grown by using a seeding machine and a compatible transplanting machine. Many annual and vegetable seedlings are propagated into "plugs" using this type of equipment. Seedling nurseries have been known to effectively halve their manpower costs by using this type of machinery, and despite the high cost of the machine, the cost benefit has been enormous.

Tray filler: Used for the filling of seedling trays. This equipment has an adjustable vibrating system and the height adjustable rotor that ensures uniform soil density as well as equal soil distribution. The filled trays are then finished and cleaned with a height adjustable, fast rotating brush unit. This equipment also has easily adjustable speed and height settings improving the productivity rate, for example machines are able to fill from 700 to 2000 average sized trays (520mm x 40mm) per hour depending on the speed setting Capacity depends on the adjustable speed setting. This equipment usually comes with options such as automatic tray dispenser, watering system, conveyor, seeding/dibble equipment.

Sprayers: Sprayers are used for applying chemical sprays (insecticides, fungicides and weedicides) or liquid fertilizers. It is advisable to use a different sprayer for each type of chemical. (Weedicide residues left from previous spray jobs can contaminate insecticides and damage plants you spray for insects).

There are many different types of sprayers available:

a) Disposable spray guns Plastic bottles with a pump action handle on top. Many chemicals are sold as a ready to use spray in this type of bottle. Refills can be bought to screw on to the spray mechanism top.

b) Pressurized back or shoulder (knapsack) pack units: Is a container which can be pressurized with a flexible hose and connecting spray nozzle. The unit is pumped by hand to raise the pressure in the container. A trigger is used to release pressure and spray as required.

c) Motorized sprayers A container which holds the chemical is pumped from by a motorized pump and out through a spray nozzle.

Sterilising Equipment: Larger nurseries mixing their own media need to use sterilising equipment to eliminate harmful bacteria but still leave beneficial organisms for healthy soil. Fan forced steam sterilisers with selected temperature and soak times are used for this purpose. The sterilisers are usually trailer mounted so that the top of the unit is removed and the trailer is then towed to the potting bench saving double handling.

Heating Trays: There is a range of commercial propagation trays available including: fibreglass, heating cable on wire mesh and flexible heat mats. Some trays are available with both heating and misting for independent use or heating only for use in humidity controlled environment such as a hot house. There are a range of sizes available or they can be custom built to suit all commercial growing needs. Misting equipment and sensors are also available for hot house use.

Handling and Care of Tools

Experience shows that many accidents are caused by the employees' improper use of tools and by the use of defective tools and equipment. Employees shall use only tools and equipment that are in good condition. Tools shall be used only for the purpose for which they were designed.

It shall be the responsibility of each employee to make frequent inspections of tools and other equipment used to make sure such tools and equipment are in good physical condition. A supervisor shall prohibit the use of any tool, device or equipment which, in his/her judgment, is unsafe.

Some of the common defects in tools and equipment that shall be eliminated to prevent accidents are the following:

- Handles which are cracked, split, broken or loose, in hammers, shovels, sledges, axes, etc.
- Mushroom heads on chisels, impact drill, etc.
- Wrenches that fit poorly, open-end wrenches, and adjustable wrenches with spread jaws, or pipe wrenches that do not hold.
- Ladders having broken or loose rungs or cracked side pieces.
- Ladders with no rubber shoes.
- Rubber protective devices having cracked, cut, or otherwise defective, rubber.
- Sharp-edged tools shall be protected or stored in such a manner as to prevent injury to employees at all times when not in use.
- When using hand tools, an employee shall place himself/herself in such a position that injury will be avoided if the tool slips.
- A machinist's (ball peen) hammer shall not be used to drive nails; a carpenter's (claw) hammer will be used.
- Pipe wrench jaws must be clean before being used.
- Worn or dull jaws on pipe wrenches should be replaced before being used.
- All files, rasps, and other hand tools that have a sharp tang shall be equipped with approved handles.
- Tools not in use shall be placed where they will not create a tripping or stumbling hazards, even during construction or overhaul periods.
- Shovels, picks, or digging bars shall be stored in a manner not to create stumbling hazards. They shall be cleaned of accumulated clay, mud, etc., before storing.
- Cutting and digging tools are the safest to use when they are kept sharp and in good condition. They shall be stored so they will not be a hazard to a person accidentally coming in contact with them.

- Pointed tools shall never be carried edged or point up in a worker's packet. They should be carried in a toolbox, carrying belt, pouch, or in the hand, points and cutting edges away from the body.
- Tools should be handed from one worker to another, never thrown. Edged or pointed tools should be passed with the handle toward the receiver.

Care of tools

- Tools and equipment shall be kept in proper operating condition and used only for the purpose for which they were designed. If proper and safe tools are unavailable, this should be reported to the supervisor.
- All tools should be inspected at regular intervals, and any tool that develops defects while in use shall be taken from service, tagged and not used again until restored to proper working condition.
- Hammers and similar tools shall be kept in good condition and shall not be used if the handles are loose, cracked, or splintered.
- Wrenches must be kept in good condition. Defective wrenches such as open-end, box-end, socket, and adjustable wrenches with spread jaws, or pipe wrenches with dull teeth, might slip.

List of useful tools and equipment

Axes	Balance (grams)	Balance (kg)
Bar, crow	Barrows, wheel	Hammer, nails
Boots (rubber)	Cans, watering	Containers, measuring
Boxes, plant transport	Brooms	Buckets, plastic
Containers, seed storage	Containers, for germination	Funnels, plastic
Cutlasses	Cutters, wire	Drums,
Files, sharpening	First aid kits	Forks, digging
Containers, for loading seedlings	Ground cover, polypropylene	Benches, wood or metal (for sitting)
Hoes	Hoses	Irrigation system
Knives, hand pruning	Knives, budding	Labels, plastic
Lamps	Lens, hand	Levels, carpenter's
Machine, calculating	Masks, respiratory	Mattocks
Sprayers, knapsack	Pencils, drawing	Pencils, marking
Pens	Pipe repair kit	Pliers, cutting
Rakes, garden	Refrigerator	Rose heads, watering cans
Rule, carpenter's	Saw, bow	Saw, hack
Scissors	Screwdrivers	Secateurs
Shovels, shredder, biomass	Sharpeners, pencil	Shears, root pruning
Spanners, adjustable and fixed	Nozzles, heavy and mist spray	Sticks, height measuring
Sieve	Tables, potting	Tanks, water
Tape	Tape, budding, grafting	

❑❑❑

3

Choosing Nursery Site

A nursery site must be located with the realization that a perfect site does not exist and that choice of site will require compromise. However, careful attention to the selection of a permanent nursery site will amply repay all the effort expended. First, a list of possible sites should be screened by the person or group wishing to establish a nursery in a given market or use area. Because selecting and establishing a permanent nursery requires a large capital investment, a team approach for final selection is probably best. The team should be composed of at least three of the following:

- Experienced nursery manager
- Reforestation specialist, silviculturist, or other potential customer
- Soils specialist
- Forest/Agriculture pathologist
- Civil engineer
- Soil Conservation Service representative
- Entomologist

Team members must be capable of blending their varied backgrounds and individual areas of expertise together, and all should have a physical sense, or "feel," for the land. Because team input is so diversified, it is far less likely that the site selected will be difficult or impossible to manage. The selection team should first review criteria for all potential sites using a site checklist (table. 1); this form may be modified to ensure that all selection criteria are listed and properly emphasized. Then each site should be visited and its merits and drawbacks discussed. Finally, the selection team should meet after all site visits have been completed to make the final selection. The entire selection process should be carefully documented in a written report.

Site-selection Criteria

The most important factors to be considered in selecting a good nursery site are discussed in the following sections.

a) Land availability and cost

Are the sites under consideration actually for sale and within the price range given to the selection team? What are the owners' sale stipulations? Look at **total** developed cost. Unimproved land may initially cost less but require such large capital outlays for development that ultimate total cost may be more. Land that may initially cost more, on the other hand, may be developed to the point that few subsequent improvements are needed, and total cost may be less.

Table 1. A nursery site checklist (recommended for use by the site selection team)

POTENTIAL NURSERY SITE

Name of the Area ....

Date ....

Location ....

Soil Survey Type ....

Top Soil

1. Texture Assessment ....
2. Depth ....
3. pH ....

Subsoil

1. Texture Assessment ....
2. Depth ....
3. pH ....

Depth of Water Table ....

Drainage ....

Water Supply

1. Adjacent Creek ....

2. Well ....

3. Other ....

4. Rise or Fall (Feet) to Supply ....

5. pH ....

Topography

1. (a) Level ....

(b) Rolling ....

(c) Irregular ....

2. Slope: .... to .... Degrees

3. Unsuitable (i.e., more than 6° or very irregular) ....

Cover

1. Main Tree/Crop Species ....

2. Common Members of Ground Flora ....

Climatic Data (Use Nearest Meterological Station)

1. Total Annual Precipitation ....

2. Highest Temperature and Month Recorded Last

Year Approximate Acreage ....

Access .... Ownership ....

b) Water

Securing an adequate supply of domestic and irrigation water can be a major problem. Water rights must be obtained for any water source. Therefore, special consideration must be given to a site where the quantity and quality of water are adequate for current and possible future requirements.

All water needs and the timing of those needs must be considered. For example, in most nurseries, irrigation is necessary during the growing season and for frost protection. Restrictions on flow and on periods of delivery must be closely scrutinized. We have to looks for the following things:

- Is the water source reliable during drought years?

- Can breaks in canals, pipelines, and other delivery systems be expected?
- What are the time frames for repair?
- Are backup sources available in emergency situations?
- Is domestic water available through a city, village, or other municipality?
- Are there rest rictions on quantity?
- Are costs high?
- Is it feasible to develop an on-site water source?
- Is water quality high?
- Are there any potential delivery problems?
- If no water is available near the site, can a transmission line be constructed?

Irrigation-water sources: Lakes are a good source of irrigation water. Storage capacity, draw- down, other uses, and contaminants must be examined before any commitment is made. Screening may be necessary to remove water-borne debris.

Streams are sometimes used for nursery irrigation and must be checked for water rights, other uses, and quality. In addition, attention must be paid to intakes, diversions for pumping stations, protection during runoff periods, and maintenance of the stream channel to ensure maximum carrying capacity. Stream water may need to be screened to alleviate contamination by vegetation, weed seeds, frogs, fish, algae, and other water-borne debris.

Irrigation water delivered through open ditches is usually controlled by irrigation districts and is subject to specific short delivery periods. Such a source is not reliable unless storage is made available on site and therefore is not recommended.

Water drawn from wells is probably one of the best irrigation sources for most locations. Draw-down and pumping capacity must be checked to ensure that water is available in reliable quantities when it is required.

Domestic or irrigation pipelines are reliable. In many instances, clean water will be supplied with adequate pressure and volume to eliminate the need for pumping. The two types of pipelines are similar, both generally well designed and constructed, although domestic water lines usually have more connections creating a high demand for water and more concern for failure of the system. Systems must be reviewed to ensure that maintenance is adequate and repairs are timely.

Water quality: Chemical contaminants may be introduced into an irrigation source through the soil or from precipitation or surface runoff. Contamination by minerals such as calcium or boron, for example, will usually be found in well water. However, because streams, lakes, and ditches also may have mineral contaminants, any potential site must have its water sources evaluated for mineral content and concentration.

Water originating from any open source (lake, stream, or ditch) is subject to contamination by weed seeds. High concentrations of these can lead to unwanted vegetation in seedbeds and cover crops-a major problem. Special, well designed screening devices can alleviate this problem.

Water-borne diseases can infect root systems and foliage. If pathogens such as *Phytophthora,* a fungus causing root disease, are present, chemical water treatment may be necessary. It is necessary to measure the pH which should be between 5.5 to 7.0. Water with pH greater than 7 favours attacks of 'damping off ' fungi in the seed beds and tends to raise the pH of the soil which in turn can reduce the growth of seedlings.

Topography

The area for nursery beds should be level, or nearly so. A slight slope (2% maximum) is beneficial for better surface drainage, but slopes greater than 2% can cause erosion, necessitating expensive control measures, and may cause undesirable translocation of soluble fertilizer salts. Furthermore, all mechanical equipment used in nurseries operates best on level ground. Moderate slopes and small rough areas maybe levelled by terracing and grading, but these operations can be expensive and usually expose infertile subsoil that can cause future seedling growth problems. Fertile topsoil should be removed and stockpiled before any major soil-moving operation is attempted and evenly redistributed afterward over the levelled area.

The importance of aspect will depend on nursery site latitude and altitude. In most localities in the temperate zone, eastern and southeastern aspects should be avoided because of greater frost danger and southern and southwestern aspects because of excessive dryness during periods of drought. Where irrigation is available, southern aspects in northern latitudes at high elevations are best because of their greater warmth. For most sites, though, a northwestern aspect is best because vegetative growth starts later in spring and is not subjected to injury by frost. Water loss through evaporation from the soil surface is not so rapid on northwestern aspects.

Topographic undulations can cause water to accumulate. Standing water, no matter how little, causes complete destruction of nursery stock because of oxygen depletion or buildup of toxic gases. Irregular topography complicates installation of irrigation systems, causes irrigation-line leakage, and makes it difficult to operate nursery equipment. Damage from early fall or late spring frosts can be catastrophic to growing seedlings. Frost hollows, which occur wherever cold air can accumulate in valley bottoms and large topographic depressions, especially where trees bar cold-air drainage must be avoided.

Location

The nursery should have a central location as near to the planting site or demand centre as possible to avoid the transportation of seedlings over long distances. In order to reduce the problems of transport, food and accommodation, and to encourage better supervision, most forest nurseries are located close to or within population centres. This avoids the need to construct dwellings and living facilities for nursery workers thus reducing costs. Sites with exposure to excessive wind should be avoided as well as valleys and old water courses that may be liable to flooding.

Preparation of the Site

As soon as the site has been selected, work can start to convert it into a more or less level area if the topography permits.

Removal of tree and vegetation cover: The first step in the preparation of a permanent nursery is to eliminate all the unwanted vegetation including trees, shrubs and small plants. It is also a convenient time to fell or save those trees either inside or outside the boundary which may either interfere with or help in future operations. On the other hand, mature trees, particularly those of the same species being grown in the nursery which are a potential source of fungal and insect pests and those which give unnecessary shade should be removed.

Removal of top soil: Following site clearing operations, all the top soil should be removed from the site before levelling to avoid muddy conditions during wet weather and after prolonged watering. It is always useful to save the top soil for later use in the potting soil mixture or for the production of compost. Avoid creating erosion problems, pollution of nearby streams and water logging conditions. Many sites either have shallow top soil or very little, if any, so it may not be necessary to remove any soil before levelling. Terraces should be constructed on slopes of more than 4 to 5 percent.

Erosion control and wind damage: Exposed slopes and the ridges of terraces should be grassed as soon as possible and during the dry season if necessary since irrigation water should be available. If there is a threat of wind damage, a wind-break should be planted around the perimeter of the nursery. A low hedge could also be considered around key sections to further reduce wind, control or prevent wind borne weed seeds and damage from dust. The species used as a wind-break and in the hedge should be relatively free of disease and insect attacks. The species to be grown in the nursery should not be used as a wind-break since they could harbour injurious pests to which the young seedlings may be susceptible.

Slower growing hedge plants are preferred, particularly those that withstand heavy pruning and maintain a tidy appearance. Any disease or insect outbreak in wind-breaks should be controlled quickly to reduce the risk of spreading to nearby nursery seedlings. Since most tall trees tend to lose their lower branches after a few years, a practical hedge can be made either by planting a row of trees with one or two rows of a low growing species (see Figure 1); or by planting three rows of trees using a species that coppices freely. Every 2 to 3 years, depending on its rate of growth, one of the outer rows is cut back and the coppice regrowth will provide the required low hedge protection. This type of hedge provides extra fuelwood as well as raw material for making compost.

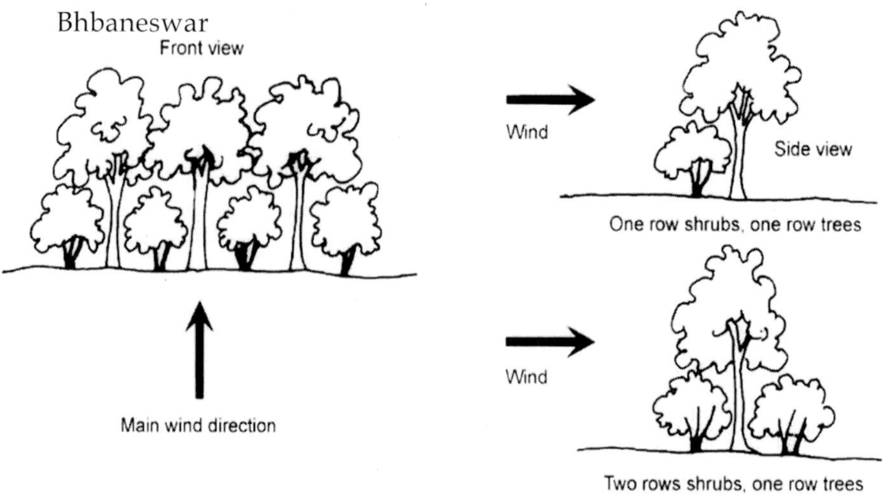

Fig. 1 windbreak pattern (A row of tree with one/two rows of low growing species)

Soil

Adequate sources of good quality soil are required in the preparation of the potting mixture and to a lesser extent for seed beds. One of the major components of the recurrent cost in a nursery with container grown plants is the provision of suitable soil. Obviously, the nearer the nursery is to the source of this material, the less will be the cost. A high demand for sand in soil mixtures has developed due to the availability and use of heavy soils. Sand particles however, do not hold moisture and they tend to be inert, conferring few, if any, benefits to the plant. This over-reliance on sand could be avoided with adequate supplies of top soil and a ready supply of compost. The most important soil properties that affect plant production are:

Texture: Sandy loams or loamy sands with good drainage are excellent for nurseries. Light soils can be worked in weather conditions too wet for heavier soils an important consideration in the Northwest. The content of clay and silt in the soil should be within 15 to 25%.

Depth: The top 4 feet of soil should be free of claypan, hardpan, shale, iron concretions, calcareous substrata, or mottled gley layers. Without artificial drainage, this depth seems a reasonable minimum; where artificial drainage has been installed, however, a minimum clear soil depth of 2 feet is probably acceptable. The top 18 inches of soil should be free of stones, which are expensive to remove, make the soil difficult to work, and interfere with nursery cultural practices.

Soil pH: The optimum soil reaction, or pH, for most tree species is between pH 5.0 and 6.0. Soils of lower pH may have fewer available nutrients, whereas soils of higher pH encourage the invasion of fungus diseases. Soil pH can be altered with soil additives such as sulfur or by injecting phosphoric or sulfuric acid into irrigation water.

Proximity to Customers, Labor, and Services

Proximity of the nursery to seedling customers, work force, transportation, utilities, and facilities for people all should be evaluated by the site-selection team. Locating the site geographically close to seedling customers seems to be most judicious, although, with the advent of transportation systems and refrigerated trucks, this is not as paramount as it once was. often, other criteria prevail.

Labour force: The number of workers needed varies widely, depending on size of the nursery, extent of mechanization, amount of work contracted out, degree of chemical weed control, and type of stock grown. The peak period of employment occurs during the short 2 to 9-week processing season, when about 1 person is required for every 65,000 seedlings processed (approx.). About 10% of this work is supervisory and administrative. Both male and female workers are needed; typically, 50 to 60% of the work force is women. Certain specialized positions such as tractor operator and irrigation specialist require some previous agricultural experience.

Transportation: A good transportation network is essential. Rail, truck, bus, or plane can be used to transport seedlings, but refrigerated transportation equipment is mandatory. County or state roads that are well travelled, maintained, and connected to freeways will aid the transport of both seedlings and people; easy connections for seedling customers and nursery administrators expedite travel and reduce transportation and per diem costs.

Utilities and fossil fuels: Telephone, electric power, and other utilities required for nursery operation must be already available or easily secured. The history of these utilities as well as their current cost, supply, and reliability must be evaluated.

❑❑❑

4

Design and Layout

An intensive soil survey and an accurate topographic map of the nursery site provide the basis for a detailed development plan. Primary points to be considered in making the plan are: size and shape of the production area, amount of clearing and land levelling needed, drainage and road network, layout of the irrigation system, fences and windbreaks and location and types of buildings. In general, the average permanent nursery should be designed to accommodate the administration, operations and the agreed production areas.

Many commercial nurseries begin as small operations with little thought given to initial or future layout design. Facilities, beds, and fields are added on an as-needed basis and their design is often directed by the availability of space rather than long-term layout considerations. Although optimization of layout and design is best achieved when planning a new nursery, modifications of an existing nursery can also be a good investment.

Sustainability can be approached through a multitude of efforts. One objective that produces meaningful results is the optimization of the nursery layout and operational design. A well-planned layout and design can greatly increase efficiency by reducing unproductive movement of people and materials. In most nurseries, labour is the single largest (about 25 percent to 35 percent) production cost, and movement of plants and materials utilizes a majority of that labour. In addition to increased efficiencies, nursery layout and design can also address the potential for minimizing the environmental impact of operations.

Planning for Labor Efficiency

Labour is an expensive but necessary component of any nursery operation. Employees complete numerous essential activities such as mixing media, transplanting, pruning plants, loading shipments, or helping customers. How-ever, the time employees spend walking, searching for an item, or waiting on someone else to complete a task also contributes significantly to labour costs but does not increase product value. Activities or resources that add costs but do not add value to the product are considered "waste" and should be limited as much as possible through proper planning and continual improvement of operating processes. A hard look at the operating processes will reveal that many seemingly necessary activities do not directly increase the customer's value of the product. With some ingenuity, even those once necessary activities can sometimes be reduced or eliminated without compromising product quality.

However, every nursery manager tackles the subject as he/she examines the operations in efforts to increase efficiency. It is easier to look at an existing system and determine the time it takes for individual activities, but is more difficult to predict time and motion requirements when designing a new nursery. Observing and examining other nurseries while planning a nursery can be very valuable. In the context, time and motion efficiencies will focus on the location of activities in relation to each other in order to reduce wasted motion.

Administration Area

The inclusion of an office and storage facilities depends on the size and useful life of the nursery and the availability of funds. Generally the office and the storage area occupy the same building with the exception of fuels and other inflammable or toxic materials for which separate storage facilities should be arranged. The office should be located close to the main entrance to avoid routine movement of personnel, visitors and vehicles from disrupting nursery activities. It would be an advantage if the location overlooked the production areas to facilitate management and supervision.

Very often washing and toilet facilities and a lunch room are overlooked when nurseries are being planned. These are essential to ensure high standards of personal hygiene as most workers are in daily contact with soil and chemicals and sometimes have to work at very close quarters with each other in large numbers.

Functional Areas For Defined Activities

Certain activities are naturally grouped into distinct functional spaces such as an office, greenhouse, or shipping area. Grouping activities into functional areas is more intuitive for some activities than others, but efficiencies can be gained by reviewing the activities that will be conducted within each defined space. The first step in defining necessary functional spaces involves determining which of the listed activities use common equipment or resources. If these activities are done by the same person or group, then designing multi-functional space makes sense. Otherwise, consider the spaces needed for individual activities and determine how those activities relate to each other. Through studying the required activities during the design process, their functional relationships should become more obvious. Functional areas may include the entrance, nursery office, sales area, propagation area, production area, media preparation and storage area, receiving area, shipping area, service area for equipment storage and maintenance, employee facilities, etc. Below are descriptions of some common functional areas.

Entrance

The organization and appearance of a nursery gives visitors and customers an impression of the operation, which directly influences sales, even for wholesale nurseries. The nursery entrance provides the first and most important opportunity to present a good image. The entrance should provide easy access to the nursery office and shipping areas, with clear signage for each. Often the office, service and shipping areas are close to the main road for easy access and to reduce the installation cost of utilities. However, the lay of the land, adjacent property use patterns and the shape of the property may dictate locating the office and other service areas some distance from the main public road. Signage becomes even more critical in such situations. Drivers of incoming vehicles should be able to find the loading areas without delay. Customer parking must be provided near the office, and the nursery entrance should be land-scaped with an attractive, uncluttered arrangement of plants, including those sold by the nursery. The entrance planting could also contain any special plant materials that need to be introduced or emphasized for marketing purposes.

Administrative site

The administrative site includes administrative offices; storage areas for equipment, trees, seed, pesticides, other chemicals, and fuels; shops; a fuel-dispensing station; an employee center; and seedling-processing facilities. The type, number, and location of required buildings can be determined with the team approach. Other administrative development could include employee-enrichment areas, holding areas for irrigation water or soil amendments, a culled-seedling disposal area, and an area for holding scrap material and used equipment until sale is possible. Although possible future expansion must always be kept in mind, the administrative complex must optimize the use of space to avoid being spread out. The results of poor or inadequate planning will cause the manager and staff considerable anxiety in future years.

Sales Area

Locating a sales area close to the office with a display of representative sailable plant materials enables customers to view plant material without travelling through the nursery. This saves time for customers and sales personnel and prevents exposure of customers to hazards.

Propagation Area

Unless all transplants are purchased from other sources, the propagation area is the heart of nursery operations and should be located in an area accessible to the production and potting areas. A propagation area located close to the office helps in communication between the office staff and the propagation managers who must make long-range decisions regarding the number of specific plants to be produced. The size and design of the propagation area are determined by production type, number of plants and species produced, and markets.

The propagation area may contain greenhouse structures of various designs. Steel frame, gutter-connected, and ridge-and-furrow greenhouses are typically larger than greenhouses with conduit and PVC frames. Conduit and PVC greenhouses cost considerably less but have a shorter life than steel frame and ridge-and-furrow greenhouses. Smaller greenhouses may be used to segregate plant species that require different rooting environments. However, several small greenhouses require more land than one or two larger houses of equivalent square footage, and this should be considered if less than ample land is available for the propagation area and facilities.

Certain plant species may be propagated outdoors in small containers or raised ground beds and do not require special propagation structures, except for possible overwintering requirements (e.g. - junipers). Because of repeated mist cycles or frequent watering, outdoor propagation areas must be located on land not normally subjected to surface water drainage issues and with well-drained soil. Predominant wind direc-tion and speed should be considered when locating outdoor mist systems so adequate wind breaks can be provided. Seeds may also be germinated in outdoor beds, although structures built to accommodate tiers or racks of seed-germinating flats use space more efficiently. Disadvantages of outdoor propagation include a lack of protection from heavy rainfall and animals such as deer and squirrels. Heavy rains may pack the rooting media, destroying aeration and contribute to soil borne diseases.

The amount of land available for propagation may determine if plant stock blocks are maintained to supply cuttings. Limited land availability may require taking cuttings from sailable nursery plants. Cutting preparation areas may be included in the layout of the propagation area. A protective structure allows for cutting preparation during inclement weather and benefits the nursery producing large numbers of plant species propagated during the winter.

Media Preparation and Storage and Potting

Proper formulation and storage of growing media for specific production systems are critical for successful crop production. Media mixing and potting may be accomplished at one central location where potting media or media components are stored in bulk quantities. Potting media or components are stored either in loose piles or in open bins often constructed of concrete. Media components are usually mixed by commercially available soil mixers or by manure spreaders or front-end loaders that scoop and dump the media several times on a concrete slab. The scoop and dump process may not adequately mix media components. Some nurseries purchase one component of the mix that already contains more precisely incorporated amendments such as micro-nutrients. The selected approach to purchasing, mixing and storage of growing media largely determines the required size and arrangement of the media preparation and storage area.

Production systems requiring long-term storage of media or media components should plan for moisture management and minimization of wind-blown weed seed. Media must remain moist, either through the use of irrigation or exposure to natural precipitation. However, excess moisture can cause leaching of incorporated fertilizer. Walls and permeable coverings such as shade cloth help minimize wind-blown weed seed while allowing rain to reach the media. Special care in monitoring and managing storage of media with incorporated fertilizer is a must.

Potting machines and motorized media mixing and transporting systems should be covered by a structure that houses a permanent potting area for the nursery. If commercial soil mixers and potting machines are not used, the potting area may or may not be sheltered. Nurseries that do not use a potting machine usually erect a permanent V-shaped hopper from which media falls onto a potting bench.

In some cases, plants are only potted up once then moved to the growing area and later sold. In other scenarios, plants may be sequenced to increasingly larger containers before being sold. In situations that require additional transplanting, potting areas located in or near the growing areas are especially beneficial for reducing unnecessary movement of plant material during each transplanting process. This can be achieved by establishing two or more potting areas throughout the nursery or using a portable potting station. Even if plants are potted only once, nurseries may use multiple media preparation and potting areas to reduce the distance travelled when placing newly potted plants in the field.

Mobile transplanting systems still require motion to replenish media and containers at the transplanting site. In other words, either the potted plants or the media and containers must be transported. The most efficient approach depends on several factors, including the shape of the land, the possibility of remote potting and storage facilities, and ease of transporting potted plant material.

Production Areas

Production efficiency is achieved by locating each of the following major activities in a defined area of the nursery.

Germination Section

This area which is set aside for the germination of seeds is best located near the office, to enable the supervisor to maintain a close watch on the activity. It is customary to sow the seeds in elevated metal or plastic trays erected waist high above the ground level. Round wood is commonly used but a more permanent foundation is recommended to avoid frequent replacement. One alternative is to build permanent seedbeds of concrete blocks with an internal measurement of 1 metre wide. Ten metres is the standard length but this varies with location and production targets. The beds may be subdivided into 1 metre quadrants to facilitate sowing smaller quantities of seeds. Since dogs, birds, rodents and insects have been known to damage seedlings during this stage, the germination section may need to be fenced and a 'roll on' cover provided for use when required.

Transplanting Area

The nursery beds for growing bare rooted plants or for standing containers occupies the greater part of the nursery area and it is here that the transplanted or direct sown seedlings are grown until they are ready for planting in the field. Normal bed width is 1 to 1.2 metres to facilitate hand tending. Length is also important and beds should not be longer than 20 metres in order to facilitate moving from one bed to the next. Bare root production beds should be located where the soil has the best chemical and physical properties.

Roads and Waterways

Efficient movement of plants, equip-ment, and employees throughout the nursery requires a well-designed road system. Number and size of roads and walkways vary depending upon equipment used and type of production system. When farm tractors and other vehicles are used, the perimeter roads should be about 30 feet wide to allow turns from narrower roads between plant beds or rows. Roads should be crowned or sloped to one side and surfaced with gravel or other hard materials to support equipment during wet periods. Firm road surfaces also prevent splashing of mud and debris onto plants. Associating waterways with roads often results in optimal land use efficiency.

Drainage ditches, spillways, grassed waterway and water runoff containment areas must also be designed to handle heavy rainfall and any irrigation runoff. The amount of water and the rate of water movement expected from normal rainfall and irrigation determine the necessary specification of these features to minimize erosion and other impacts. Most container bed surfaces are minimally absorbent of surface water which increases the amount and speed of runoff. Fast moving surface water can result in significant erosion, to a point that limits access. Although not currently required by regulation in most states, surface water drainage systems in container production should contain all surface water on the nursery site in ponds, lakes and/or wetlands. At times this is not possible due to the rate and amount of rainfall, but at least normal rain events should be contained as well as any usual runoff from irrigation. Regulations in some states require that the first inch of rainfall be captured on nursery property.

The location of ditches, grassed water-ways, and holding ponds must be established in relation to the topography of the site. With some hilly sites and depending upon the nature of the entire watershed, multiple containment facilities may be necessary. Contained water may be used for irrigation or released following treatment. Treatment could include such approaches as vegetated waterways, settlement ponds, constructed wetlands, etc. The design of surface water drainage and containment systems is critical and should be performed by a professional.

Service Area

Equipment storage and repair facilities, along with pesticide, petroleum, and fertilizer storage facilities, comprise the nursery service area. The service area is often located close to the nursery office due to the cost efficiency of locating electrical and sewer services close together. This may create other issues such as having work areas adjacent to more public areas, etc. The service area also needs to be accessible to supply trucks and or tractor servicing these facilities. The type of equipment and supplies needing shelter or storage determines the size and type of facilities. Enclosed metal buildings are excellent for repair and maintenance shops and may be used for storage of small pieces of equipment such as hand sprayers, chain saws and lawn mowers. Equipment repairs by commercial mechanical businesses may be practical if timely and affordable service can be assured. Nurseries that use this approach reduce their requirements for equipment repair facilities.

Pesticide storage facilities should be separated from other activities and have secured access. Mixing is usually done adjacent to the pesticide storage area to minimize transporting concentrated pesticides. A pesticide storage building must be properly designed and identified as containing poisons. Local, state, and federal requirements for pesticide storage and mixing facilities should be determined before proceeding as those regulations can change frequently.

Employee Facilities

Employee facilities are usually located adjacent to the office or service areas but should not be close to the pesticide storage and mixing area. Depending upon the season and amount and tim-ing of customer traffic, employees with dirty boots, etc. in the office area could interfere with marketing activities. Restrooms, showers, personal lockers, refrigerators and dining tables are usually provided for employees. Employee parking between the service area and the employee facilities is a convenient arrangement. In large nurseries or for nurseries with non-contiguous production areas, employee facilities may best be provided in multiple locations. This would reduce travel time for employee breaks and increase employee comfort. Secured material and tool storage may also be part of more remote employee support facilities.

Shipping Area

The need for shipping areas, as well as their size and location, varies greatly from nursery to nursery and depends upon such things as the size and type of the operation, proximity to production areas, diversity of plant materials, size of trucks, type of customers, etc. The cost of a plant increases each time it is moved. However, it may be more cost effective to gather plants in a central location for shipment. Some nurseries load plants directly from production areas, but most operations have designated loading areas within the nursery where plants are placed prior to shipment. Relatively large nurseries shipping primarily in tractor-trailers generally require a shipping area. Production areas are usually not accessible by such trucks, and loads are often comprised of a wide array of plant materials for multiple customers. In such cases, plants can be collected and grouped according to shipments at a shipping area. Placing plants in this area before customer or freight truck arrival reduces loading time, but irrigation, and may be even shade, must be provided. Nurseries selling primarily to local landscapers who pick up plants with smaller trucks and trailers often have designated pick-up locations in production or holding areas. Possible conflicts between customer vehicles and production activities should be considered when establishing pickup locations.

□□□

5

Nursery Operations

Nursery operations involves various activities such as, seed sourcing, Seed bed preparation, Sowing seeds, Potting, Pricking Out, Shading, Watering, Weeding, root pruning, application of additional fertilizers or manure.

Sourcing Seed and Pre-treatments

Sourcing tree seed: It is important to try and use good quality seed in planting. Seed quality is important for growing high quality seedlings. Seed purity, weight, germination potential, and vigor must be accurately assessed so that the correct sowing rate can be calculated and an evenly spaced seedbed attained. The need for stratification and the treatment time should be carefully determined for seed of different species and geographic origin because it can affect germination rate, vigor and amount and seedbed uniformity. Seed can be collected from trees locally from farms, forest or public land. It is a good idea to collect seed with neighbours; then bulk this seed together and share it out. In this way, diversity is maintained in planted material, which is important in promoting good performance. When getting seed from a supplier, it is important to look at the seed and check that it appears to be of good quality, and hasn't been collected too early (is immature) or contains many empty seed. Before planting a lot of seed, it is a good idea to first check it's viability by seeing if it germinates well. Once you have an idea about viability, then it will be possible to estimate how many seed need to be planted to get a certain number of trees.

Seeds can be bought from local seed dealers, from NGOs, and from institutions and Department. More information on how to source seed can be obtained by getting in contact with your local seed centre, or by reading resources.

Pre-treating seed: Sometimes it is important to treat seed before it is planted, in order to improve on the level, speed and uniformity of germination. These treatment methods can be used when seed does not otherwise germinate well.

The most common methods for pre-treatment are:

a) Soaking seed in hot water until the seeds look swollen. This is used for seeds of trees such as tamarind, most acacias, leucaena, albizia etc.

   Procedure: boil water and pour it over seeds in a container. Allow to cool and leave the seed in the water until the seeds look swollen.

b) Soaking seed in cold/cool water. This method is recommended for seeds that have soft seed coats such as sesbania, dalbergia species, gmelina, Gliricidia, sesbania etc. The time for soaking varies between 12-48 hours, depending on the tree species.

   Procedure

   - Soak the seeds in cold water which is 2 times its volume.
   - Remove all floating seeds.
   - Sow the remaining seeds at the bottom in containers in the nursery or direct sow directly in the field after.

c) Cracking the seed shell method: This method is used for tree species with a hard coat like melia, *croton megalocarpus* etc. The cracking is done to allow water penetration for easy germination. Cracking is done using a sharp knife, a stone or a cracking machine.

d) Nicking-cut slightly the seed at one tip to allow water penetrate.

Fine and light seeds such as Eucalyptus, Casuarina are sown in transplant beds and later pricked out into containers. It is important that fine seed is mixed with sand and uniformly broadcast on the seedbed to avoid overcrowding that can lead to damping off. Do not sow the seed too deep in the soil; the depth of holes should depend on the size of the seeds (usually 5 mm to 1 cm). Sowing too deep is likely to prolong seed germination period or seeds may lot. Put a seed in each hole; gently cover with soil equal to the size of the seed itself. Water the pots. Make a light shade with netting material to cover the pots or the seed bed after sowing.

Seed Bed Preparation

Nursery beds can be arranged in different ways. Potted seedlings can be raised on a flat bed, or can be set into a sunken bed, which is a basin like excavation of about 1 m by 1 m and about 10 cm deep. Such a structure holds seedlings together, and helps to conserve water in dry areas. Raised beds are used for establishing bare-rooted seedlings; as the sides of the bed can be broken down to reveal the roots of plants, ready for transplanting. Staking slats of wood into the ground in a square or rectangle with sides of about 1 m and then filling this structure with soil (mixed with sand if possible) makes a raised bed. Alternatively, the sides can be made from bricks or wood planket. It is normal to germinate small seeds such as e.g. eucalyptus, sesbania etc. in such seedbeds, to allow them to germinate before transferring small seedlings to pots or planting them directly.

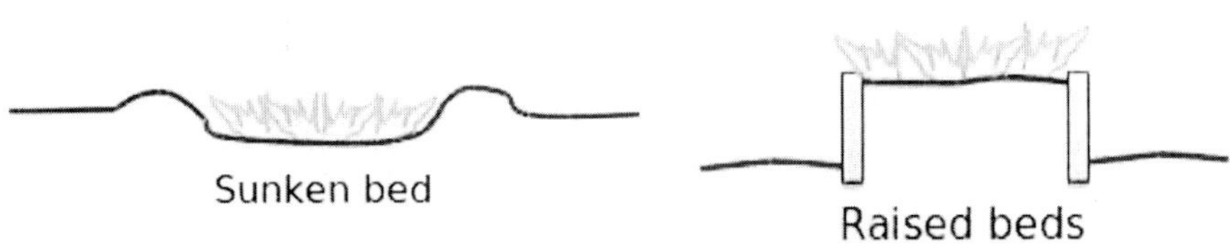

An example of a sunken bed and raised bed

Potting

Potting mixture (soil, sand and compost/manure) should be moistened and then pressed into containers to a depth of about three-quarters of the height of pots. Pots should then be topped up more loosely with mixture and pressed down lightly to about 2 cm below the top. Heavy compaction should be avoided at the top of pots because it will inhibit root penetration.

Before planting seed, containers should be watered lightly. Sometimes, more than one seed can be planted in a pot and then, if more than one germinate, seedlings can be removed to leave a single individual. This approach might be used if germination rate is expected to be quite low.

Sowing Seed

When raising seedlings in pots it is normal to use polythene tubes that are around 10 cm in diameter and 20 cm deep, though the size of pots will depend on the species in question and the time that seedlings will be in the nursery. Instead of using plastic tubes, other locally available materials that can be used include small tins, milk packets, cardboard boxes, banana fibre containers, calabashes and clay pots. It is better to use open-bottomed than closed containers, since this allows healthier root development and possible root pruning. If tins or other containers are being used, it is important that holes are made in the bottom to allow the movement of water. Seed sowing time depends on the species and the time it takes to attain size for planting out (30-45 cm). It is important the seeds are sown in time to enable the seedlings attain the recommended size (at least 30-45 cm in height). This should be attained before the onset of rains.

Pricking out

This is the process of transferring young and tender seedlings from seedbeds into containers (pots) Fig 1. Pricking out should be carried out when the seedlings reach a height of 2 cm. This is usually about two weeks after sowing but depends on the species. The pricking process:

- Water the seedbed and containers properly before commencing the operation.
- Ensure adequate shade is available.
- Take an empty container and fill with water to ¾ levels.
- Hold the leaves of the seedlings and insert a pencil thick stick (dibble) underneath the root system to loosen the soil.
- Pull out the seedlings gently and immediately put them in the container with water. Note that if the roots of the seedlings are kept under sunshine they lose water and may die.
- Make a hole at the center of the pot using a stick.
- If the roots are too long clip off the tip.
- Do not hold the stem of the seedling because they are tender and feeble this may injure the seedlings.
- Hold the stick in the tilting position and insert it in the soil about one centimeter away from the seedling to the same depth as the hole.

- Push the soil towards the seedling to hold it tightly. This ensures that all the air pockets around the roots are closed.
- Using your fingers cover the hole you made.
- Water the containers properly once more after planting. Seedlings pricked out from same batch of the seed bed should be arranged in the same place.

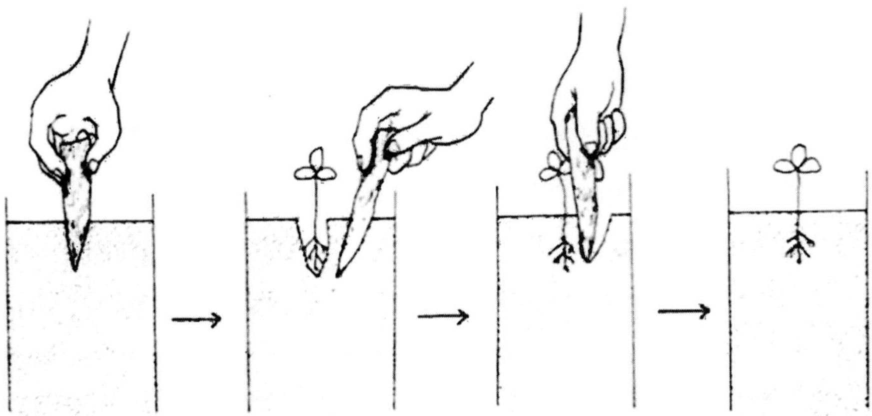

Fig.1. Process of transferring of young seedling into container

Shading

Construct a shade to protect the seedlings from direct sunlight for two to three weeks after pricking out. Use locally available materials such as bamboo, grass, mats, or banana fibers for shade construction.

Watering

The regular supply of clean water is essential to plant growth. Plants are made out of more than 90% water. When grown in containers, nursery plants have only a limited volume of substrate and do not have the ability of mature trees to search for water from below the soil surface. The amount of water seedlings require depends on upon the following point.

- ***Seedling age:*** More water is required after germination when the seedling is young and at pricking out but this requirement reduces as the seedling grows in age. The amount of water should be reduced four weeks before the seedlings are planted out. At that stage, the soil can be left to dry out completely and the plants to wilt for a day. The process should be repeated several times.

- ***Amount of sunlight.*** If the area is sunny, more water is needed and viceversa. However, do not keep the area shady for too long to reduce water use.
- ***Soil type.*** A sandy soil loses water faster than a soil with high clay content hence need more frequent watering. However a clayish soil becomes hard and cracks if it dries out.

The substrate should be watered thoroughly so the water should be directed to the soil and not the leaves except to dust the soil lightly. A watering can or a hosepipe with a nozzle should be used to ensure uniform distribution water and one should water the whole bed and not just the plants in the centre of the bed. Low water pressure is good but one should ensure that water gets to the bottom of the container to avoid a dry and hard bottom, which will affect the growth of the roots as they get to the bottom of the container.

Water should be clean to ensure seedling health so water from such sources as kitchen waste should not be used. Too much water can damage the plants just as much as not enough water because of water logging which makes the roots not breathe. Avoid direct use of hosepipes while watering the seedlings as this may wash away the soil, Use a watering can or an empty tin with holes at the bottom.

A tribal Woman watering the nursery by using a tin water cane

This is used to reduce the power of water drops which cause soil erosion. It helps distribute water evenly. As already mentioned, watering in principle should be done twice a day early in the morning and late in the afternoon after 4.00 pm. when the sun is not strong. If this is not possible, then water once a day in the evening depending on the crop. Avoid under watering and over watering. Use adequate amount of water.

Weeding

Weeds are a threat to healthy seedlings development. They compete with seedlings for nutrients, water and light hence they must be controlled. With you hands or a dibble gently pullout unwanted growth (rouging) this should be done whenever weeds are observed. Remove all the weeds around the beds and don't leave any debris around unless you are sure that this can be converted to compost.

Application of Additional Fertilizers (Nutrition)

Fertilizers or manure is applied when the seedlings show sign of weakness. The most common fertilizers are NPK. It's important to note that manure from livestock can also be used. However weakness of the seedlings can be caused not only by insufficiency of the nutrient but also by pests and disease.

Root Pruning

Root pruning is the cutting of roots to control root system development beyond the container. Why root prune? When seedlings have reached to a certain size and their roots become longer than the depth of the pots. If the roots are left without pruning, they penetrate into the ground and develop the root systems there. Once the root system develops under the ground, it is hard to move the pots, and if the roots are cut when the seedling is old, the seedlings will be weakened; hence periodical root pruning is required before the root system reaches into the ground. The period and interval of pruning depends on different species and other conditions. Root pruning should be done regularly preferably every 2 -3 weeks. Prune when seedlings are the height of the span of your palm and when their roots have started to penetrate into the under surface.

Root pruning using a knife (the knife should face downwards)

Procedure of root pruning are as follows:

- Water the seedlings properly before root pruning.
- Using a sharp knife or wire or scissors to cut the long roots underneath the container. You can also uplift the containers (wrenching) to cut overgrown roots.
- Water the seedlings well after root pruning. This helps the plant withstand moisture stress.
- Note that to reduce root pruning, you can Place the seedlings on a bed of stones or on polythene sheet and this reduces root development.
- If the seedlings are in a raised bed, prune the roots by using a knife or wire underneath the bed, soon after watering.

Hardening off

Hardening up is to expose the seedlings to harsh conditions to make them strong so that they will be able to survive under harsh climate in the field after planting out. It is also a gradual preparation of seedlings for field conditions.

Hardenings up process are as follows:

- When the seedlings grow and reach the planting size, the shade should be removed to exposure to more sunshine.

- Reduction in watering intensity (quantity) and frequency of water twice a week and later once a week.
- Before planting out, root pruning should be carried out frequently or re-arrangement of pots to allow more adoption to stress.
- Good preparation for out planting results in good field survival, therefore hardening off should be done 2-3 weeks before out planting time.

Postponing Planting

If it's not possible to plant when the seedlings are ready for planting out (reached right size for planting-1.5 ft) or the seedling are not bought; cut the tips of the plant to suppress further growth so that they will not be overgrown during next season planting.

6

Producing Plants from Seed

This method is otherwise called as sexual method of propagation. Sexual reproduction refers to multiplication of plants by seeds. Seeds are formed after successful pollination and fertilization by the union of male and female gametes. Meiosis division takes place in the course of fusion and the chromosome numbers are reduced to half, which after fertilization becomes normal. The plants raised through seed are called seedling plants. Sexual propagation involves careful management of germination conditions and facilities and knowledge of the requirements of individual kind of seeds.

Success of seed propagation depends upon fulfilling the following conditions

A. Using seed of proper genetic characteristics to produce the cultivar or species, of provenance desired. This can be accomplished by obtaining seed from a reliable source or dealer.

B. Using good quality seeds which germinate rapidly and vigorously to withstand possible adverse environmental conditions in the seed bed and provide a high percentage of usable seedlings.

C. Manipulating the seed dormancy by applying pre-germination treatments or proper timing of planting.

D. Providing proper environment for seed germination i.e., supplying sufficient water, proper temperature, adequate oxygen and either light or darkness (depending upon kind of seed) to the seeds and resulting seedlings until they are well established.

Advantages

1. For evolution of new varieties through breeding, the hybrids are raised from seed.

2. This is the most popular, simple and easy method of propagation.
3. Seed propagation is only mean of diversity particularly in the selection of chance seedlings.
4. Seedling plants are long lived, productive and have greater tolerance to adverse soil and climatic conditions and diseases.
5. Seed propagated rootstocks are hardy and develop better root system.
6. Viruses don't transmit through seeds, thus mostly the seedlings are free from virus diseases.
7. Occurrence of polyembryony (more than one embryo in seed) in citrus and mango leads to the development of uniform seedlings as in asexual method.
8. Seed is the source for production of rootstocks for asexual propagation

Disadvantages

1. Seedlings have a long juvenile period (6-10 years) and come into bearing later as compared to asexually raised plants.
2. Due to segregation of characters, the progeny is not true-to-type.
3. It is not economical to handle larger trees, as less number of trees can be accommodated per unit area and the cultural operations are difficult.
4. Seeds of many fruits are to be sown immediately after extraction from the fruits as they lose their viability very soon e.g. cashew nut, *jamun*, jackfruit, citrus, mango and papaya.
5. The beneficial influences of rootstocks on scion variety cannot be exploited in sexual propagation.
6. Seedling plants usually produce fruits inferior quality.

The following are the most common methods of raising tree seedlings from seed and there are many variations in these methods:

- In an open bed from which seedling plants are lifted and planted with roots bare of soil.

- In containers, either singly or in multi-cavity trays which are taken to the planting site and the seedlings planted with a ball or plug of soil around the roots.
- Transplanting natural regeneration seedlings often found under mature trees into containers.

The production of seedlings in the nursery is best described under the following stages of development and growth:

Stage 1: Seed Handling

Stage 2: The Germination Process

Stage 3: Transplanting the Young Seedlings

Stage 4: Tending the Seedlings

Stage 1: Seed Handling

Seed Procurement and Storage

Which species and provenance to use is pre-determined by Head Office who also have the responsibility for collecting or purchasing national or local seed requirements. When the seed arrives it is therefore normally ready for sowing. It may be necessary however to store the seed for short periods. Use rodent and insect proof containers which should be airtight if possible. Avoid extremes of temperature and humidity and fluctuating temperatures. Keep the seed in a refrigerator if one is available, placing the seed in a cloth bag and then placing the cloth bag in a plastic or glass container. The cloth bag will absorb the condensation inside the container during periods when the electricity supply is disrupted. Do not store seed in a freezing compartment. Keep the seed at a uniform temperature in the refrigerator above 5^0 C. Since fluctuations in temperature can be harmful, divide the seed lot on arrival into manageable units and remove from storage only the amount required to be sown on the appointed day. Apply seed fungicide and insecticide on arrival of seed at the nursery if this was not previously done. Some seeds cannot be stored for long periods, eg, neem, Spanish elm, mahogany. When these seeds arrive at the nursery they should be given priority for immediate sowing.

Good Practices of Fruit Collection, Seed Extraction and Storage

- Fruit/pod/seed collected directly from trees is of high quality because they are healthy and not been exposed to soil moisture or soil microorganisms. Selecting seed from many parts of the crown, assures broad genetic base of the seed collected.

- If seed extraction and processing can not be conducted immediately, place sacks or containers of seed in a dry and cool room with good air circulation. Place the sacks or containers on a shelf or rack to facilitate air movement. Sacks may also be hung from the ceiling.
- Mature fruits should be processed for seed extraction. Fruits that are undeveloped or infected with insects and disease should be discarded. Proper extraction methods should be conducted carefully to avoid damaging seed. Normally, 2-3 days of drying is enough. Rubbing and crushing the fruits will expedite seed extraction.
- Some common tools used during extraction include sacks, tarps, trays, buckets, tanks and sandpaper (or other abrasive material). Many species have wings pod/seeds that should be removed. Most wings can be detached by crushing or rubbing the seed. Large, firmly attached wings can be removed by hand and discarded.
- If seed is to be sown immediately after extraction and cleaning, no further processing is required. However, if it will be stored, even for a short period, the seed of most species require drying in order to maintain viability during storage. Whether seed requires drying or not depends on its classification as recalcitrant or orthodox.

Broadly seeds can be classified as recalcitrant and orthodox. Recalcitrant seed must retain high moisture contents to remain viable and generally can be stored for only a few days or weeks. If stored under humid conditions, such as wrapped in a moist cloth or paper, recalcitrant seed may remain viable for a slightly longer period. Species with recalcitrant seed include *Artocarpus heterophyllus* (jackfruit), *Azadirachta indica* (neem), *Calamus* species (rattans), *Durio zibethinus* (durian), *Eusideroxylon zwageri* (ulin), *Theobroma cacao* (cacao) and many dipterocarps (*Shorea, Hopea, Palaquium,* etc.). Orthodox seed must be dried to a low moisture content before they are stored. The orthodox seed of many species can be stored for one year at room temperature or several years in cold storage with little loss of viability. Orthodox seed stores best at moisture contents of 5-8%, species with orthodox seed include *Acacia mangium* (mangium), *Eucalyptus* species,

Tectona grandis (teak), *Paraserianthes falcataria* (falcata) and *Gliricidia sepium* (gliricidia).

Table 1. Criteria for quality planting material of some important plant species

Tree	Propagation method	Quality standards
Silvicultural species		
Acacia auriculiformis	Seed	15-30 cm tall
Acacia nilotica	Seed	5 month old
Albizia lebbeck	Seed	5-6 month old
Azadirachta indica	Seed	1 yr old
Casuarina equisetifolia	Seed/nodal cutting	35-45 cm tall, 5-6 months old
Dalbergia sissoo	Seed	60 cm tall, 3-4month old
Dendrocalamus strictus	Seed	45-60 cm tall, 1 yr old
Eucalyptus spp.	Seed, two leafy cutting	30-45 cm tall, 6-8 months old
Gmelina arborea	Seed	20 cm tall, 6 month old
Prosopis cineraria	Seed	1 yr old
Prosospis juliflora	Seed	1.5 m tall, 1 yr old
Sesbania grandiflora	Seed	60 days old
Tectona grandis	Seed	1-2 cm collar dia, 3-4 months dia
Horticulture, ornamental species and vegetables		
Aonla	Budding	6-12 months old seedling
Areca nut	Seed	15-18 months old
Banana	Sucker/Tissue culture	2-3 month old seedling
Ber	Budding	1 yr old budding seedling
Black pepper	Cutting	120 days propagules
Brinjal	Seed	4-5 week old, 12-15 cm height, 6-8 leaf stage
Broccoli	Seed	4-6 week old
Cabbage/cauliflower	Seed	3-5week old seedling
Cashewnut	Epicotyl/Softwood grafting	5-6 month old propagules
Chilli	Seed	6-7 week old seedling, 15-20 cm height

Citrus	Grafting	1-2 yr old graft, 75-90 cm height
Cococnut	Seed	1-1.5 yr old, stem girth 10-20 cm
Custard apple	Seed	5-6 month old seedling
	Grafting	1 yr old graft
Fig	Cutting	8-12 month old, 4-6 buds, 30-40 cm height
	Layering	2.5-3 months old propagules
Gaillardia	Seed	3-4weeks old, 4-6 leaf stage
Gerbera	Seed	5-6week old, 2 leaf stage
	Sucker/Tissue culture	5-6 leaf stage propagules
Grape	Cutting/Grafting	15-20 cm long, 3-4 buds, 2.5 cm dia stock
Guava	Layering	6-9 month old propagules
	Tounge grafting	45 days after removal from mother plants
Jasmine	Cutting	60-65days old propagules
Marigold	Seed	3-4 weeks old, 5-6 leaf stage
Onion	Seed	6-8 week old, 20-25 cm height
Papaya	Seed	15-22.5 cm tall
Pomegranate	Air layering	20-25 cm height, 3-6 month old
Sweet orange	Budding	150 days old
Sapota	Softwood and approach grafting	60 days after removal from mother plants, leaves fully turn green
Rore	Budding	3-6 month old propagules
Tomato	Seed	3-4 week old, 12-15 cm height, 4-6 leaf stage

Seed Dormancy and Pre-Treatment

Seed dormancy refers to a state in which viable seeds fail to germinate when provided with conditions normally favourable to germination, *i.e.*, adequate moisture, appropriate temperature and light. Dormancy in seeds may be advantageous or problematic during seed handling. The advantage is that it prevents seeds from germinating during storage and normal handling. On the other hand, where dormancy is complex and seeds need a specific pre-treatment, failure to overcome these problems may result in poor germination. Some pre-treatment procedures are not directly related to seed dormancy but are carried out in order to speed up the germination process. It is important to maximise germination of all seed sown and the following are the main pre-treatment or pre-germination treatments that may be necessary.

Types of dormancy: Different types of dormancy include

1. Exogenous Dormancy

This type of dormancy is imposed by factors outside the embryo. In exogenous dormancy, the tissues enclosing the embryo can affect germination by inhibiting water uptake, providing mechanical resistance to embryo expansion and radicle emergence, modifying gaseous exchange (limit oxygen to embryo), preventing leaching of inhibitor from the embryo and supplying inhibitor to the embryo. It is of three types:

a) **Physical dormancy (seed coat dormancy)**: Seed coat or seed covering may become hard, fibrous or mucilaginous (adhesives gum) during dehydration and ripening as a result they become impermeable to water and gases, which prevents the physiological processes initiating germination This type of dormancy is very common in drupe fruits i.e. olive, peach, plum, apricot, cherry etc. (hardened endocarp), walnut and pecan nut (surrounding shell). In various plant families, such as, Leguminosae, the outer seed coat gets hardened and becomes suberized and impervious to water.

b) **Mechanical dormancy**: In some fruits seed covering restricts radicle growth, resulting in dormancy of seeds. Some seed covering structures, such as shells of walnut, pits of stone fruits and stones of olive are too strong to allow the dormant embryo to expand during germination. The water may be absorbed but the difficulty arises in the cementing material as in walnut. Germination in such seeds does not occur until and unless the seed coats are softened either by creating moist and warm conditions during storage or by microbial activity.

c) **Chemical dormancy**: In seeds of some fruits chemicals that accumulate in fruit and seed covering tissues during development and remain with the seed after harvest. It is quite common in fleshy fruits or fruits whose seeds remain in juice as in citrus, cucurbits, stone fruits, pear, grapes and tomatoes. Some of the substances associated with inhibition are various phenols, coumarin and abscisic acid. These substances can strongly inhibit seed germination.

2. Endogenous dormancy

This type of dormancy is imposed by rudimentary or undeveloped embryo at the time of ripening or maturity. This can be of different types such as morphological, physiological, double dormancy and secondary dormancy.

A. **Morphological dormancy (Rudimentary and linear embryo)**: Dormancy occurs in some seeds in which the embryo is not fully developed at the time of seed dissemination. Such seeds do not germinate, if planted immediately after harvesting. Plants with rudimentary embryos produce seeds with little more than a pro-embryo embedded in a massive endosperm at the time of fruit maturation. Enlargement of the embryo occurs after the seeds have imbibed water but, before germination begins.

B. **Physiological dormancy**

a) **Non-deep physiological dormancy:** After ripening time is required for seeds in dry storage to lose dormancy. This type of dormancy is often transitory and disappears during dry storage. Temperate fruits such as apple, pear, cherry, peach, plum and apricot, cultivated cereals, vegetables and flower crops, have this type of physiological dormancy which may last for one to six months and disappears with dry storage.

b) **Photo dormancy**: Seeds that either require light or dark condition to germinate are termed as photo-dormant seeds. It is due to photo-chemically reactive pigment called phytochrome widely present in some plants. When imbibed seeds are exposed to red light (660-760 nm), the phytochrome changes to red form (Pfr), thereby substituting the germination process. However, when seeds are exposed to far-red light (760-800), Pfr is changed to Pf which inhibits germination process.

c) **Thermo dormancy:** Some seeds have specific temperature requirement for their germination, otherwise they remain dormant. Such seeds are called as thermo dormant .For example seeds of lettuce, celery and pansy do not germinate if the temperature is below 25^0C.

Physiological dormancy is of 3 types

I. **Intermediate physiological dormancy**: The seeds of some species require a specific period of one-to-three months of chilling, while in an imbibed and aerated state, commonly called as moist chilling. For example, most of temperate fruit seeds require moist chilling to overcome seed dormancy. This requirement led to the standardization of world famous, horticultural practice of stratification. In this process, the seeds are placed between layers of moist sand in boxes and exposed to chilling temperatures (2 to 7^0C) for the period varying from 3-6 months to overcome dormancy.

II. **Deep physiological dormancy:** Seeds, which usually require a relatively long (>8 weeks) period of moist chilling stratification to relieve dormancy as in peach.

III. **Epicotyl dormancy**: Seeds having separate dormancy conditions for the radicle hypocotyl and epicotyl, is called as epicotyl dormancy e.g. *Lilium, Hepatica antiloba and trillium.*

C. **Double dormancy**

In some species, seeds have dormancy due to hard seed coats and dormant embryos. For instance, some tree legumes seed coats are impervious and at the same time their embryo are also dormant. Such seeds require two years for breaking of dormancy in nature. In the first spring, the microorganisms act upon the seed making it weak and soft and then embryo dormancy is broken by chilling temperature in the winter next year. Combination of two or more types of dormancy is known as „double dormancy. It can be morpho-physiological i.e.

combination of under developed embryo and physiological dormancy or exo-endodormancy i.e. combination of exogenous and endogenous dormancy conditions i.e. hard seed coat (physical plus intermediate physiological dormancy).

D. **Secondary dormancy**

Secondary dormancy is due to germination conditions. It is a further adaptation to prevent germination of an imbibed seed if other environmental conditions are not favorable. These conditions can include unfavorably high or low temperature, prolonged darkness and water stress. It is of two types:

I. **Thermo dormancy**: High temperature induced dormancy.

II. **Conditional dormancy**: Change in ability to germinate related to time of the year.

Separation of empty seed: This usually applies to small seed and may be affected by soaking the seed in water overnight and discarding those that float as this often indicates emptiness. If the seed is extremely valuable or of uncertain viability, those that float can be broadcast sown to gain experience and confirm results. Soaking for one or two days is also used to encourage early, even germination.

Pre-treatment to weaken seed coat or break dormancy: Aside from emptiness, many seeds exhibit dormant characteristics. For example, the hard seed coat of many legume species prevent the uptake of moisture. Some seeds contain chemical germination inhibitors which must first be broken down whilst others have partially developed embryos which need to mature. The techniques used to counter this problem vary with the particular species and include the following measures:

- Many woody plant species require a cool temperature and moist media treatment for seed germination to occur. Cool-moist treatment of seed to promote germination is called a **stratification** treatment. In most cases seeds with a stratification requirement, such as teak, need a minimum of a 6 to 12 week treatment in a moist, aerated environment followed by a warm, moist environment for seed germination to occur. Seed must be maintained cool and moist since either warm temperatures alone or dry media during the stratification process will inhibit germination. Burying the seed between layers of wet, sifted sand is a form of stratification that facilitates germination of teak seed.

- The **scarification method** facilitates germination by physically removing enough of the seed coat by nicking, piercing, chipping, filing, or drilling with the aid of a knife or needle or by burning. Although effective with tough seed coat species, this method is rarely used because it is time consuming and impractical where large seed quantities are involved.

- **Hot water** overcomes dormancy in Leguminoseae by creating tension which causes cracking or weakening of the seed coat. The method is most effective when seeds are submerged into the hot water and not heated together with the water. A quick dip is also better to avoid heat damage to the embryo. For most hard-coated *Acacia* species, it is enough to bring the water to the boil and place the seed in the water after removing the flame. The seed can be planted as soon as the water cools or kept in storage for short periods.

- Soaking hard-coated seed in **sulphuric acid** for periods varying from 5 to 60 minutes, depending on the concentration of the sulphuric acid, can improve germination but care and experience is necessary as results vary considerably between species and seed lots. The duration of acid treatment should aim at reaching a balance in which the seed coat is sufficiently ruptured to permit the seed to imbibe water but without the acid itself reaching the embryo. The acid may be re-used several times although its strength will gradually decline. After soaking, the seed is removed from the acid and rinsed under running water for at least 10 minutes.

- Pre-treatment with **fungicides or insect repellent** is useful if there is a risk of insect or damping-off fungal attack. The method is to dust the seed with an appropriate fungicide/insecticide prior to sowing. Special pretreatment against rats, mice, birds or small insects may be necessary in which case Head Office should be consulted to help obtain a suitable repellent.

Stage 2: The Germination Process

This stage begins with the absorption of water and ends with the elongation of the radicle (root) and involves the following steps:

Step 1 - Seed coat imbibes/absorbs water

Step 2 - Water moves into the inner membrane

Step 3 - Stored food reserves (protein and starch) activated

Step 4 - Cells multiply and elongation begins

Step 5 - Seed coat cracks

Step 6 - Radicle emerges and seed has germinated.

Time of Sowing

Time of sowing is important since seedlings must be of the right size and quality by the start of either the short rains which normally peaks in May or the start of the longer rains in September/October. If it takes 3 months from sowing to be ready for planting, then seed should be sown this amount of time in advance of plants being needed. Since all the plants will not be needed at the same time, a certain amount of staggering may be necessary, based on the demand which should have been planned or estimated in advance of sowing.

Method of Sowing

The two basic methods of sowing seed for container or bare root production are broadcast and direct sowing. The choice of method will depend on the local conditions and experience. It should be noted that:

- The yield of plantable seedlings from a given seed lot is usually much greater with broadcast sowing.
- Development of transplants are sometimes slow compared with undisturbed direct sown seedlings.
- There is a risk of root distortion during transplanting with broadcast sowing.

When seed is plentiful, of good quality and its cost is only a small proportion of total nursery costs, direct sowing is sometimes preferred. Direct sowing is also easier if the seed is large enough in size to be handled individually and the risk of root distortion is reduced with this method.

Broadcast Sowing

This is usually done in specially constructed beds containing sand or sandy loam or pre-mixed sterile sowing medium in a wood or concrete block frame or metal tray. Clay or heavy soil must not be used as it is essential to have free drainage in the beds. Make the surface smooth, level and firm but not compact. The seed should not be broadcast too thickly over the bed surface to avoid overcrowding and to allow each seedling to have sufficient growing space. A 'roll on' cover may be necessary to protect the seed from birds and to provide shade.

Broadcast Methods

For tiny seeds such as Eucalyptus species, the practice is to mix the seed intimately with an equal part of fine, dry sand of a similar size and spread the mixture evenly with the fingers. An alternative method is to make a 20 by 20 cm tray using strong mosquito or similar wire netting. The tray is then covered with fine sand particles just large enough that they will not penetrate the netting. On top of this the seed is added, usually sufficient to sow 1 square metre of the germinating seed bed. After shaking the tray, the small seed will find the openings and be deposited on the bed evenly and in the quantity desired. If the seed is mixed with sand of a similar size, it will be sown more evenly and uniformly. The work can be simplified if seedbeds are divided into 1 metre quadrants so that the recommended amount of seed to be sown per square metre can be premeasured.

Direct Sowing

The advantages of direct sowing are lower cost and the avoidance of damage to seedlings through careless transplanting. Although only one seed is necessary if the germination rate is high, the aim is to sow an average of two to three seeds per container. With small seed, special methods need to be used to regulate the amount to be sown. In some overseas nurseries, a shaker is used, made from a small bottle with graduated holes in the lid set to allow a given number of seeds to drop per shake. Good results have been obtained from direct sowing *Acacia auriculiformis, A. mangium* and *Azadiracta indica* in pre-filled polythene bags. The use of cell packs that resemble egg boxes is also popular since the germinated seedling can be lifted with a plug of potting medium thus avoiding shock on transplanting.

Depth of Sowing

A general rule is to cover the seed to a depth equal to twice the seed diameter and not deeper than 1 cm in the case of *A. mangium* and similar small seed sizes. The recommended practice is to apply the lightest cover which is capable of withstanding routine watering. The tendency in some nurseries has been to sow the seed much deeper than necessary to avoid the washing out or uncovering after heavy watering. In such cases it is the water that should be regulated, using nozzles and pressure that will reduce the force of the water.

Care of Seedbed and Direct Sown Container

After sowing, cover the seed with a layer of fine sand or sifted nursery soil or potting mix to the desired thickness and water the bed or container lightly. Do not press the seed into the seedbed or container. Shade the seedbed from direct sunlight immediately after sowing with jute bags, newspaper, bamboo or any suitable available cover. This may be placed directly on the seedbed at first as a source of protection and to maintain humidity and an even damp moisture condition that favours germination. The cover must be lifted to about 30 cm or higher at the first sign of germination and removed completely as soon as possible afterwards to avoid weak, pale looking seedlings. Watering must be done gently with a fine spray and in large nurseries where piped water is available, mist nozzles are recommended for watering seedlings, using a filter to avoid sediment from clogging the small openings of the mist nozzle.

Stage 3: Transplanting the Young Seedlings

Depending on the species, seedlings are normally ready for transplanting from 3 to 5 weeks after germination when 2 or 3 pairs of leaves have formed. Some selection may be necessary at this stage since all the seeds do not germinate at the same time. The important points to remember when transplanting are:

- Seedlings to be transplanted should be fresh and firm. This is achieved by watering them well the day before they are to be taken out from the seedbed. At the time of lifting the soil should be only moist, not wet.
- In lifting the seedlings take care not to damage the roots or the stem. Do not pull them individually. Lift the seedlings by raising a whole block of soil using a thin piece of wood and free a bunch of seedlings together. Afterwards, the soil should fall away easily as there should be few lateral roots at that stage to retain it.

- Hold the seedlings only by the leaves or by the 'match head' or needles in the case of the pine. If the stem or root is held, the plant is more susceptible to damping off fungi from bruising, crushing or infection by the fingers.
- The roots must not be exposed to air or sunlight for longer than a few seconds. As soon as the seedlings are lifted, place the roots in water in a shallow container to avoid drying out before transplanting.
- Plant the seedlings at the same depth as they were in the seedbed or perhaps a little lower but never higher.
- Fill the container and prepare the transplanting space using a broad flat stick or your fingers rather than the customary pointed stick that makes a narrow conical shaped hole that encourages abnormal root development.
- Cut off the tip of the root if it is too long using a sharp knife, making a clean cut, leaving about 5 cm of root.
- Firm the soil gently around the root. Do this carefully to avoid air pockets at the bottom or side of the slot. Do not use force as this could bruise the seedling and reduce the porosity of the soil medium by compaction.
- Plant the seedling in a straight, upright position in the centre of the container. The roots should not be allowed to bend or curl in the planting space (see Figure 1).
- After transplanting a batch of seedlings, water them again immediately with a fine sprinkle and place them under shade if the days are particularly hot. This should not be necessary if the transplants were sufficiently 'hardened off' or grown in full sunlight.

Experienced nursery workers will gradually reduce the frequency and amount of water during the last 10 days before transplanting in order to harden the seedlings, particularly if they are more than 4 to 5 weeks old. The above principles apply equally to the transplanting of seedlings from seed beds or from pots in which more than one 'direct sown' seed has germinated.

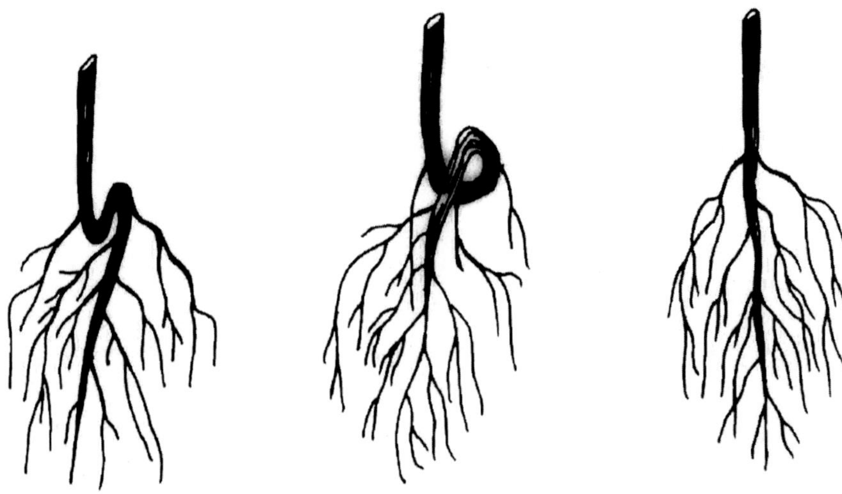

Fig.1 Different root systems of seedling due to bad potting/transplanting practices

Light and Shade

Light affects all the different stages of growth of container tree seedlings and should be managed by the grower to increase photosynthesis which takes place in the foliage. With photosynthesis, the plant is able to produce the basic building blocks of its life such as carbohydrates, amino acids and fats and also generate oxygen which is essential for the respiration of all organisms. Water is absorbed in the roots and transported to the leaves where it is combined with carbon dioxide in the presence of light to produce sugar. The products of photosynthesis are transported throughout the plant and are used in respiration to release the energy required for a wide variety of growth and maintenance functions.

For the above reason, seedlings will do best under high light intensities and the goal should be to grow seedlings in the open wherever possible. Trees high enough to provide ample walking space for adults may be necessary to provide shade for nursery workers during the transplanting operation. In contrast, low shade spread on supports of 30 to 50 cm above the ground is sometimes necessary to provide temporary shade for seed beds and transplant bags in periods of intense heat or heavy rainfall. It is a mistake to use shade once the seedlings are over the shock of transplanting although nursery staff will be tempted to keep them under shade to reduce the amount of watering needed and to mask the effect of bad transplanting. Plants that have been transplanted will require shade for only up to 3 days if the job has been satisfactorily done.

Transplanting Natural Regeneration Seedlings

Natural regeneration, sometimes called wildlings, refers to seedlings that have germinated below parent trees either individually or under forest cover. In recent years, there has been an increased demand and use of seedlings from this source due to:

- Shortage of planting stock
- Inadequate seed supply
- Seed germination failure
- Higher demand at short notice

In the absence of proper care and attention to detail, the survival of newly transplanted wildlings in the nursery and in the field is often sporadic. Good results can be obtained if the following practices are followed:

1. Select younger seedlings before tap roots have developed and penetrated deep into the soil.
2. Arrive on the site with the necessary tools and equipment consisting of:
   - Fork or pick-axe to loosen the earth
   - Burlap or plastic for packaging
   - Coir dust or compost material to reduce moisture loss from roots and stem
   - Sharp knife or secateur for root pruning
   - Container with supply of water to keep package damp and reduce shock
3. Do not wrench seedlings which are firmly rooted in the soil.
4. Place lifted seedlings in water; root prune if necessary.
5. Package seedlings in bundles of 20 or less packing damp coir dust or compost around the roots.

6. Transplant seedlings in containers at the nursery with minimum delay.
7. Temporary shade may be required until roots recover and start absorbing moisture.

Stage 4: Tending the Seedlings

Watering

Germination Beds and Transplants

The single most important factor in germination and seedling production is water but too much water can be just as harmful as too little water. With seeds and tiny seedlings, it is not necessary to provide heavy doses of water as this not only leaches out the soil nutrients but can expose seed or wash out seed before germination begins. The answer to this problem is not to plant the seed any deeper, as is sometimes done, but to adjust the hose nozzles to allow a finer spray or cover the seed temporarily until they germinate. Water the containers before sowing and immediately after covering the seed with the sowing medium or sifted sand. The sowing medium should not be allowed to get dry. Several light applications of water at this stage is far better than one or two heavy applications. Keep the soil moist but never sodden, watering preferably in the mornings and avoiding the mid-day period when the sun will cause excessive evaporation. Water deposited on the surface of leaves heats up readily in the overhead sunlight, causing the tender leaves to scorch.

Over Watering

Excess water is nearly always damaging since the water tends to replace the air in the soil and cause compaction, which in turn restricts the respiration process of the plant. Further, excess water promotes development of fungal diseases like damping off. The visible symptoms of over watering are slight to severe yellowing and stunted growth. This usually occurs near spray lines and at drip points where the sprinkler is attached to upright pipes. It is also common in areas where, due to depressions on the nursery bed, water is allowed to stand in pools. Over watering tends to occur in nurseries with heavy soils. Sometimes large groups of seedlings exhibit 'wave' formations where the watering system is not supplying water uniformly to all the plants.

This condition should not be confused with blocks of plants with stunted growth which is normally caused by a deficiency of nutrients.

Under Watering

Wilting is one of the early signs of under watering. In sandy or heavy soils, there is a tendency to apply insufficient water, wetting only the superficial layer of the container. For this reason, it is often necessary to sample some bags, particularly those at the edge of beds to see if they have in fact received an adequate supply. Any signs of wilting should be immediately tended to by the addition of water so as to prevent permanent damage.

Watering Methods

Watering may be either by hand or by irrigation. Hand watering with cans fitted with a rose spray or knapsack mist nozzles are the obvious methods for small nurseries. With these methods an adequate number of filling points must be provided. Hand watering with a plastic or rubber hose fitted with a rose spray is also popular in small nurseries where water pressure can be provided. This method can of course work with large nurseries but it suffers from the same disadvantages as the other hand methods in that uniformity of coverage is difficult to achieve.

Surface irrigation was used but the method is no longer widely used. Great skill is required in laying out such a system which is also not an efficient method since it wastes water that is often in short supply.

The ideal system for large nurseries is overhead sprinkler irrigation as it is easily controlled and provides the most uniform method for the application of water. A large number of systems are available and the heavy drops from some systems have been known to cause nursery workers to direct sow seeds much deeper than normal resulting in large-scale germination failures. On the other hand, some rotary systems can be so efficient that the need for shading transplants is often unnecessary. Whenever a sprinkler system is being installed in a new nursery, it is advisable to obtain expert advice before a final decision is made.

Weed Control

Weeds compete with the seedlings for nutrients, water and light. If they are not removed in time, this competition will suppress the young plants because the weeds are usually more vigorous and grow at a faster rate. The most trouble, some are grasses or dicotyledonous plants that grow from a root stock. If such a weed is cut off at the ground level, it will sprout again and continue to grow from the carbohydrates stored in its root tissue hence the need to remove the whole plant.

As much as possible one should prevent the presence and dispersal of weed seeds in the nursery. Most weeds produce large quantities of seed which are easily transported by water, wind and also brought in by introduced top soil, chicken litter and farm yard manure. No weeds should therefore be allowed to flower and fruit along paths and roadways or unused land in the nursery. Grassy areas should be regularly cut and trimmed. A thick hedge around the nursery helps keep out weed seeds that are otherwise brought in by wind. Since it is more difficult to eradicate weeds after they have invaded seedlings growing in containers and in transplant beds, both the potting soil and the pre-filled containers may be watered in advance so that the germinated weeds can be removed in advance of transplanting. For this purpose, containers should be filled up to 4 weeks in advance of transplanting or direct sowing operations if weed free potting soil is not available.

Pruning

Ideally, the roots of most high quality tree seedlings should be straight, carrot like or compact with many healthy succulent fibrous roots and root hairs. The corresponding stems should also be straight and sturdy with a symmetrical dense crown demonstrating a good balance between shoot and root. Because of the many variables involved, achieving this goal is not an easy task and one of the main limitations is the small size of the container or the comparatively unlimited space in the case of bare root production. Some time after transplanting, the roots of most seedlings will tend to exit the pot through the drainage holes at the bottom or sides of the container into the soil if there are no ground cover barriers. If they fail to penetrate the container a vigorous root will develop other abnormalities by having to run circles inside the pot. In situations where the root exits the pot, root pruning will be necessary.

Some growers overlook this routine operation and wait until just before the plants are about to be lifted and sent to the field. Although giving the appearance of a healthy seedling in the nursery bed, the wrenched plant with obvious unbalanced root to shoot ratio becomes the first casualty in the field. This is because the vigorous shoot loses water through its leaves and soon wilts as the incapacitated root is unable to perform its function of replacing the loss by absorbing water from the soil. The condition of the root system and in particular that of its absorbing surface, the hair rootlets through which water and nutrients are taken up, determines to a large extent whether a seedling will survive field planting or not. Pruning stimulates root growth and causes the roots to become compact and fibrous rather than long and thin.

As a general rule, carry out the first root pruning as soon as the roots appear through the bag and enter the ground. Repeat the pruning every two weeks or as necessary. Do not wait for the last moment. Carry out the operation at least three weeks before the plants leave the nursery for the field. This is to allow sufficient time for the plant to replace its loss of roots by developing new laterals under the more favourable nursery conditions.

Since root pruning reduces the water absorbing surface of the root system, it is best carried out on a cloudy day when transpiration is low. If there are signs of wilting, then the plants should be watered. Root pruning can be done by either lifting each container or by cutting the tap root of bare rooted stock.

Root Pruning by Lifting

In this method (see Figure 2), the first row of plants are carefully removed to provide operational space and each plant in the second row is lifted and placed in the vacant row after the roots growing out of the container have been cut with pruning scissors or a sharp knife. Do not shock the plants by lifting them forcibly as this damages the roots which do not heal easily or properly and the laterals tend to develop slowly. Using force can also destroy the intimate relationship between the roots and the soil in the bag and the benefits of growing the seedling in a container will have been lost.

Root Pruning in Bare Root Transplant Beds

While the seedlings are in the transplant beds, the roots are cut underground with a sharpened spade. The pruning level should not be too deep nor too shallow ranging from 10 to 30 cm below the soil surface depending on the species and the size of the seedling. Using a well sharpened spade, vertical cuts are made between the seedlings to shorten long lateral roots (see Figure 3). Then the spade is again forced into the soil, this time between the row of seedlings at an angle of about 45 degrees starting as far away from the seedling as the intended pruning depth.

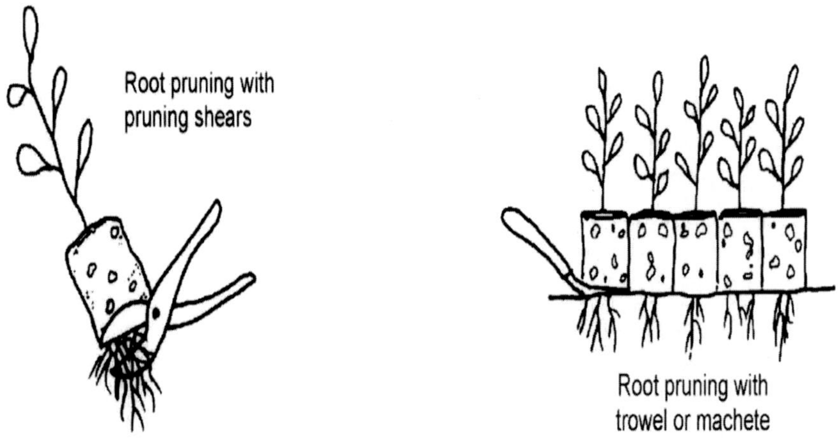

Fig. 2 Pruning container grown plants.

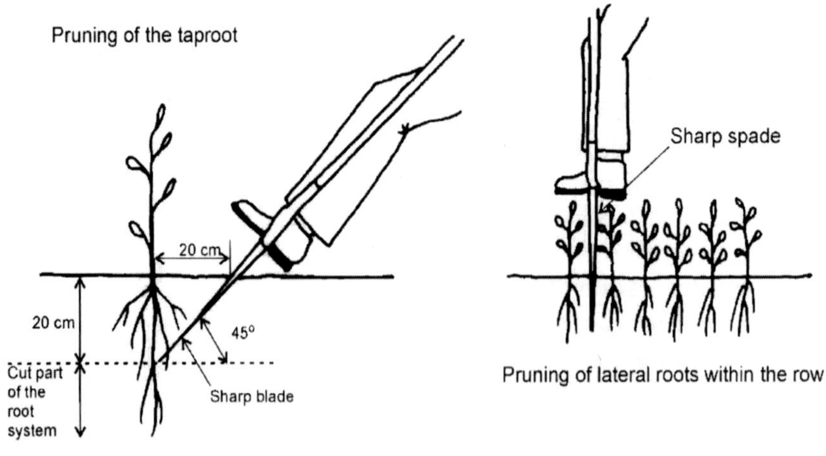

Fig. 3 Root pruning bare root seedlings in a transplant bed

Cutting Back Transplants

If necessary, most oversized seedlings can be cut back to a convenient height of 5 to15 cm and allowed to recover in the nursery before planting in the field. Cut back plants tend to be bushy for the first few months after planting in the field but a leading shoot usually comes away and well formed trees are later produced. Where possible, a second pruning of unwanted shoots can be done as soon as they appear to facilitate the growth of a single stem. This method of cutting back stems is particularly useful if plants reach the desired size sooner than required and the need exists to hold them back for two to three months. On the other hand, it is normally undesirable to cut back plants and hold them for the next planting season as, by then, the plant would have used up all the fertility in the small bags. It would also most certainly have developed abnormal roots which would, in all likelihood reduce its chances of survival in the field.

Production of Stumps

Stumps are bare rooted seedlings with most of the shoot and more or less all lateral roots pruned so that only a short piece is left. Seedlings intended for stumps are raised in beds at wide spacing with or without transplanting to attain a diameter at the root collar of 1 to 2 cm. The stem is cut back to between 5 and 10 cm above the root collar, leaving one or two pairs of buds for sprouting and developing into a new plant. If a longer part of the shoot is left, the stump may produce several sprouts. Lifting of nursery seedlings can be done by hand (digging and pulling) or by machine. After lifting, the seedlings are transported to a stumping shed. The lateral roots are trimmed off (for teak, close to the tap root). The tap root is also cut to a length of about 15 cm (see Figure 4). At low rainfall locations, planting of stumps should commence soon after the rains begin. The seedlings in the nursery should therefore be prepared into stumps before the growth flush period, to obtain a higher growth potential when planted. Therefore, stump preparation should start just before the rainy season Larger stumps can be stored up to 4 months in so called pit stores at temperatures between 25 to 30°C. Seedlings known to have the capacity to be grown as stumps include teak, neem, Honduras mahogany, *Cassia siamea*, *Gmelina arborea*, *Gliricidia*, *Spathodea*, and bamboo and growers may have experience of several others.

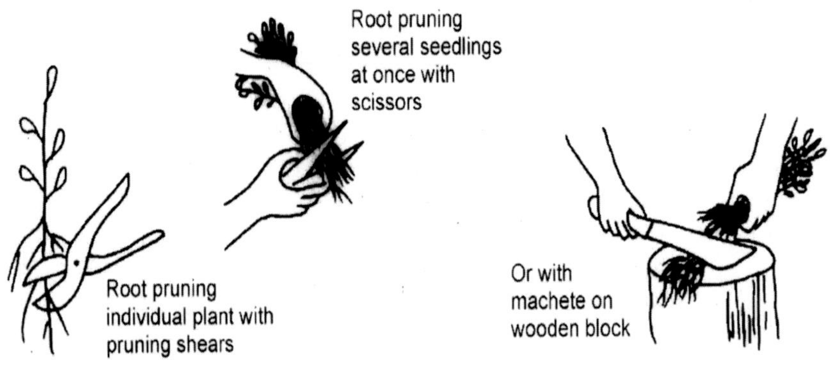

Fig. 4 Pruning bare root seedlings after lifting

Hardening Off

This refers to the progressive withdrawal of the favourable conditions in which the seedling has developed in the nursery with the objective of conditioning the plant for survival in the harsher environment in the field. This 'rougher' treatment should begin not later than halfway through the life of the seedling in the nursery. The main treatment involves the reduction of water and full exposure to sunlight. Reduce the amount and frequency of watering gradually to two or three applications per week depending on the species, the potting medium and the local climatic conditions. Hardened seedlings have the following characteristics:

- Firm, lignified stems, often brown in colour
- Sturdy, well developed crowns with leaves extending over three quarters the length of the stem
- Vigorous, healthy, leathery leaves, compact rather than oversized and weak

Nursery workers often leave the practice of hardening off to the last week or two when already the first rains have started. As a consequence, the plants are never really in a hardened condition and often suffer a set back being unable to withstand a dry spell after planting. For this reason it is advisable to extend the growing time by sowing seeds two or three weeks ahead of time and slowing down the rate of growth by reducing water during the dry spell.

During the hardening off period, the reaction of the plants must be observed closely with a kind of intuition or instinct since the progress cannot be measured and the changes in the plant do not take place abruptly but gradually. Immediately before the plants leave the nursery they should be given a good final watering. Do not lift the seedlings by the stem at this or any stage. In summary, the following treatment can be applied to facilitate the hardening off process:

- Reduce the frequency of watering
- Reduce the quantity of water
- Cut off fertiliser applications early
- Expose plants to full sunlight as soon as possible
- Cut back roots
- Cut back shoots if oversized
- Ensure that each plant has adequate space

Grading of Plants

The production of seedlings of low quality and poor condition are some of the main causes of low field survival and these should not be taken to the field for planting. Also the few survivors grow slowly and must be tended for longer periods than good quality, faster growing seedlings. Losses have to be replaced incurring additional cost and also represent a waste of time, effort and money. The following points should be considered in judging whether a plant should be sent from the nursery or destroyed.

Transport of Plants

This is the final responsibility of the nursery supervisor before the plants leave his or her care. About 4 months of hard work and skill can come to nothing if the plants are not properly prepared and given the necessary care for the journey to the planting site. The main problems are damage at the root collar, bad lifting, vibration on the way from the nursery, wind damage, drying out and sun scorch, and bad storage and holding facilities.

Seedlings should be handled as little as possible to minimise physical damage therefore avoid transporting seedlings to a temporary holding site to eliminate the need for additional handling.

The seedlings should not at any stage be lifted by holding their stem. Care must be taken with handling of seedlings at every step of the transporting process:

- From the transplant beds to a waiting area in the nursery
- From the waiting area to the trailer/truck
- From the truck to the temporary nursery site (when necessary)
- From the temporary nursery site to another truck
- From the truck to the plantation site
- From the plantation site to the planting hole

Given all these steps, without due care in handling seedlings, their survival in the field can be less than 40 percent although good quality plants were produced in the nursery and the planting in the field took place under favourable conditions. The container grown plants should be packed and moved at each stage in open top boxes until they arrive at the planting hole. It is important to pack the boxes tightly so they cannot move about in the truck.

Important Termand Terminology

Dormancy: The dormancy in seeds might be due to hard seed coat, impermeability to water and gases, physiological immaturity of embryo, deficiency of some endogenous growth promoters or excess of endogenous growth inhibitors. Different methods like Stratification, scarification, and chemical treatment are used for breaking dormancy in seed to improve germination. Stratification consists of keeping seeds at low temperature (around 7^{0}C) by placing them in alternate layers of moist sand. While, in scarification the seeds are either treated with concentrated acid (acid scarification) or the seed coat is weakened mechanically, so as to make it permeable to water and gasses. In other instances, seeds can be pre-treated with some chemicals like thiourea or potassium nitrate to improve seed germination. Soaking seeds in water also overcome seed dormancy, if it is primarily due to endogenous growth inhibitors, which leach due to prolonged soaking of seeds.

Germination: Germination refers to the emergence of a new plant from the mature seed. In other words, the activation of metabolic machinery of the embryo leading to the emergence of a new seedling plant is known as germination. For the germination to be initiated, the seed must be viable, that is, the embryo must be alive and capable of germination and the environmental conditions must be appropriate. Seed germination involves different phases of germination. These phases are characterized by a rapid increase in water uptake during the imbibation phase followed by a period where there is little water uptake called the lag phase. During the third phase, there is increase in fresh weight of seed, which leads to the emergence of roots.

Apomixis: This is a natural mechanism during which vegetative embryo is produced instead of sexual or zygotic embryo. In other words, the apomixis occurs when an embryo is produced from a single cell of the saprophyte and does not develop from fertilization of two gametes. The seedlings produced are thus true to type and vigorous as compared to those produced through sexual means. Some species or individuals produce only apomictic embryos and are know as obligate apomicts. While other produces both apomictic and zygotic embryos are known as facultative apomicts, which occur in citrus.

Polyembryony: Polyembryony, as the name indicates refers to the seeds having more than one embryo in the seed. One of the embryos arises from the union of male and female gametes and is called gametic or sexual embryo, whereas the others are produced by simple mitotic division of cells of nucellus without the help of male gamete in their formation. The phenomenon of nucellar embryo is of common occurrence in citrus and mango.

❑❑❑

7

Producing Plants Vegetatively

Vegetative reproduction is the method of producing plants without the use of seed. It is independent of sexual propagation process as there is no involvement of sex organs. Otherwise called asexual reproduction. It takes place due to mitotic division. Mitotic division continues in shoot tip, root tip and cambium. When some portion of plant is wounded, mitotic division takes place. Under mitotic division, chromosomes divide longitudinally to form two daughter cells. This forms the basis of asexual propagation. The plants raised through asexual process are identical to mother plants. Cutting, division, layering, budding grafting and micropropagation are main techniques of asexual propagation.

Advantages

1. In some fruit plants (ex. Banana, pineapple), which do not bear seeds, this is the only method of propagation.
2. The plants are generally true-to-type, uniform in growth, yielding capacity and fruit quality.
3. Have short juvenile phase, thus come into bearing earlier than seedling plants.
4. The advantages of rootstocks can be obtained by budding or grafting susceptible varieties on resistant/ tolerant rootstocks.
5. Plants have restricted growth, thus cultural practices and harvesting are easy.
6. Using asexual methods, desirable characters of a mother plant can be perpetuated/ multiplied easily.
7. The benefits of rootstocks and scion are usually exploited through asexual propagation.

8. Repairing of damaged portion of plant is possible through asexual propagation as in case of bridge grafting.
9. It is possible to convert a non-productive local variety into productive improved variety by using asexual methods.
10. It is possible to grow several varieties on one plant or change variety of existing plant by top working.

Disadvantages

1. New variety cannot be evolved by this method.
2. Such plants are not so vigorous and long-lived as the seedling trees.
3. Germplasm conservation requires lot of space and is expensive as compared to storage of seeds.

Cuttings

It is the method of propagating fruit plants in which the part of a plant (generally stem) having at least few buds, when detached from parent plant and placed under favorable conditions develop into a complete plant resembling in all characteristics to the parent from which it was taken. This method is commonly used in plants, which root easily and readily, thus, multiplication of plants is very quick and cheap. For example, fruit plants like phalsa, baramasi lemon and grapes are commercially propagated by cuttings.

The hardwood cuttings are the common method of propagation, which are prepared from fully mature tissues. Round cuttings (Plate A) are preferred over angular immature cuttings (Plate B). The shoots of about one year old or more can easily be used for preparing hardwood cuttings. In case of deciduous plants (grape, pomegranate and fig) the cuttings are made after pruning. While in evergreen plants (baramasi lemon), the cuttings can be prepared during the spring (February -March) and rainy season (August-September). Generally the cuttings of 15-20 cm length and having 3-5 buds are made. The lower cut is given in a slanting manner just below the bud to increase the absorption of nutrients. The upper cut is given at a right angle to reduce the size of the wound and as far as possible away from then upper bud to avoid its drying. After the cuttings are prepared they should be allowed to dry. These cuttings are usually tied in small bundles (20-25 cuttings) and buried in moist soil/sand for a certain period for healing of wounds, which is known as callusing.

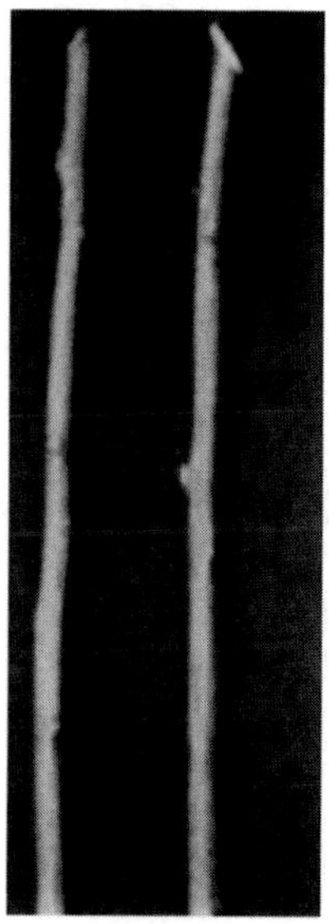

Plate A: Cuttings from evergreen plants

Plate B: Cuttings from deciduous plants

What are the conditions needed for good results from cuttings?

- High humidity
- Full light intensity
- Protection from pests and diseases
- Absence of water logging
- Absence of windy conditions
- Appropriate rooting medium

Maintaining high air humidity is perhaps the most important because cuttings lose water rapidly especially through the cut ends which have as yet no roots to take up water to replace water loss. Any water logging around the cut portion will inhibit or delay the development of roots. Death due to dessication before rooting is the major cause of the lack of success in propagation by cuttings. High humidity can be maintained by frequent manual spraying or through the use of mist propagation systems.

Importance and advantages of propagation by cuttings

- Now-a-days, propagation system are more market driven than production-driven, which means that propagators must first analyze market demands and then select and develop cultivar utilizing optimum propagation techniques to produce plants for the customers.
- Cuttings are still most important means of propagating ornamental shrub-deciduous species as well as broad and narrow leaved types of evergreen plants.
- Cuttings are also widely used in commercial green house propagation of many floricultural crops.

Advantages

- Many new plants can be produced in a limited space.
- It is inexpensive, rapid and simple and does not require the special techniques necessary in grafting, budding or micropropagation.
- No problem of graft incompatibility with rootstock and poor graft union etc.
- No variation due to variable seedling rootstocks.

Disadvantages

- The advantages of rootstock like induction of dwarfism, drought or disease resistance etc. cannot be utilized
- Plants raised through cuttings have lesser longevity as they are susceptible to various diseases and insect-pests.

Types of cuttings

On the basis of plant part used and relative positions on a plant, cuttings are classified in various groups.

A. **Stem cuttings :** A stem cutting is any cutting taken from the main shoot of a plant or any side shoot growing from the same plant or stem. The shoots with high carbohydrate content usually root better. Broadly, there are four types of stem cuttings, namely hardwood, softwood, semi-hardwood and herbaceous cuttings.

i) **Hardwood cuttings :** Cutting from mature and lignified stem of shrubs and trees are called as hardwood cuttings. Hardwood cuttings are prepared during dormant season, usually from one-year-old shoots of previous season's growth. The size of cuttings varies from 10 to 45 cm in length and 0.5 to 2.5 cm in diameter, depending upon the species. Usually, the cuttings of 25-30 cm length, with pencil thickness are preferred. Each cutting should have at least three or more buds. While preparing the cutting, a straight cut is given at the base of shoot below the node while a slanting cut, 1 to 2 cm above the bud is given at the top of cutting. However, in case of hollow pith species such as kiwifruit, top cut should also be close to bud to avoid drying up of top portion. For tropical and subtropical crops, straight cut is given at top in order to minimize transpiration loss and slant cut should be given at the base to expose more area for absorption of water and nutrients. This helps in maintaining the polarity of the shoot and if rain occurs, water does not accumulate on the tip of the cutting, which saves the cutting from fungal infection. A number of deciduous fruit plants like grape, kiwifruit, hazel nut, chest nut, fig, quince, pomegranate, mulberry, plum, olive, and gooseberry etc. are commercially propagated by hardwood cuttings.

ii) **Semi-hardwood (green wood) cuttings :** Semi-hard wood cuttings are those made from woody, broad-leaved evergreen species with partially matured wood. These types of cuttings are mostly used in evergreen fruit plants like mango, guava, lemon, jackfruit some shrubs and shrubby ornamental plants. The length of the cuttings varies from 7 to 20 cm. The cuttings are prepared by trimming the cuttings with a straight cut below the node and removing a few lower leaves. However, it is better to retain two-to-four leaves on the top of the cuttings. While planting $1/4^{th}$ cutting should be inserted in the soil. The best time for taking such cuttings is summer, when new shoots have emerged and their wood is partially matured. It is necessary that leafy cuttings should be rooted under conditions when water loss from the leaves is minimum. Commercially, such cuttings are rooted under intermittent mist, fog or under polyethylene sheets laid over the cuttings.

iii) **Softwood cuttings :** Cuttings prepared from the soft-succulent and non-lignified shoots, which are not hard or woody, are called as softwood cuttings. Such types of cuttings are very prone to desiccation. Therefore, proper arrangement for controlling humidity is required. Usually the size of cutting is 5-5.7 cm but it may vary from species-to-species. In general, some leaves should be retained with this type of cuttings. The best time for preparing softwood cuttings is late summer. Softwood cuttings generally root easier and quicker than other types, but require more attention and sophisticated equipments. Similarly, the temperature should be maintained 23 to 27^{0}C during rooting at the base of cuttings.

iv) **Herbaceous cuttings :** Herbaceous cuttings are made from succulent non-woody plants like geranium, chrysanthemum, coleus, carnation and many foliage crops. These are usually 7-15 cm long with few leaves retained at the upper end. These are rooted under the same conditions as that of softwood cuttings, requiring high relative humidity. Bottom heat is also useful for initiation of rooting process. Herbaceous cuttings of some plants exclude a sticky sap (as in geranium, pineapple, cactus etc.) that interferes with root initiation process. In such cases, basal ends of cuttings should be allowed to dry for few hours before planting. Generally, fruit plants are not propagated by herbaceous cuttings.

b) **Root cuttings :** Propagation by means of root cuttings is also a simple and cheap method of vegetative propagation in species, which are difficult-to-propagate by other methods. In general, the plants, which produce suckers freely, are easily propagated by root cuttings. For preparation of root-cuttings, roots which are of 1cm thickness and 10-15 cm long are cut into pieces. The best time for taking root cutting is late winter or early spring, when roots are well supplied with stored food material. However, in temperate fruits, root cuttings are prepared in the month of December and are kept in warm place in moss grass or wet sand for callusing and are then transplanted in the nursery during February-March in the open beds. Blackberry and raspberry are commercially propagated by this method. However, kiwifruit, breadfruit, fig, rose, mulberry, apple, pear, peach, cherry and persimmon are also propagated by root cuttings.

c) **Leaf cuttings :** Propagation through leaf bud cuttings is partially useful in species where leaves develop root system but die because of non-development of shoot system. Leaf bud cuttings are particularly useful when planting material is scarce because each node in leaf can be used as cutting. Leaf bud cutting should preferably be prepared during growing season because buds if enter into dormancy may be difficult to force to active stage, thereby inhibit the rooting in such cuttings.

d) **Leaf bud cuttings :** A leaf bud cutting consists of a leaf blade, petiole and short piece of stem with attached axiliary bud of actively growing leaves. In leaf bud cutting, 10-15 cm stem portion is used when propagating material is small. It is an useful method of propagation in blackberry, raspberry, lemon, camellia etc.

Mist Propagation

Mist propagation is a mechanical way of spraying water on to the cutting:

- To maintain high humidity
- To reduce moisture loss by evaporation and transpiration
- To keep the cutting alive until enough roots are formed to support itself advantages of mist propagation include:
- Less disease and insect problems
- Reduced respiration and transpiration
- Maximum use of light intensities
- Increase of carbohydrate reserves
- Improved root quality
- Increased percentage of cuttings that root

It is possible to use a continuous system for rooting small quantities but intermittent systems give best results. Growers should at least have a clock to turn the system on in the mornings and off in the evenings and an intermittent cycle timer connected to a solenoid valve to control the frequency of the mist. Other devices that are available for misting systems include:

- Photo cells that activate the system when the light reaches certain levels.

- An electronic leaf to maintain a uniform level of humidity on the leaf surface

Mist beds may be on the ground or raised, preferably exposed to full sunlight and fenced to avoid the action of wind blowing away the fine spray of water.

Layering

Layering is a method of vegetative propagation, in which roots are induced on the shoots while they are still attached to the mother plants. This is a alternate method of propagation in fruit plants which do not root easily when detached from the mother plants. Most commonly used methods of layering are air layering, ground and mound layering.

Advantages

- It is an effective method of propagating species that usually do not root easily by cutting as in mango, litchi, filberts and kumquat etc.
- It is a natural method of propagation in blackberries and raspberries.
- It does not require precise control on water, relative humidity or temperature, as for other methods of propagation.
- Easy-to-perform and does not require much infrastructure.

Disadvantages

- Costlier in areas where labour availability is a problem.
- Limited number of plants can be produced.
- Plants produced through layering have usually small brittle roots.
- The mortality rate is particularly higher in air layered plants.

Air Layering

Air layering or marcotting has several advantages in that large plants can be produced quickly by relatively unskilled workers and with fewer operations than are required for budding or grafting. Also they can be brought into fruit production much quicker than budded or grafted trees or plants produced from seed.

In air layering (see Figure 1), the branch to be propagated is girdled by removal of a ring of bark 25 to 38 mm in width. It is advisable to scrape the surface of the wood in the girdled area to remove the cambium, otherwise the area can heal over quickly and fail to make roots. The girdled area is then covered with a ball of sphagnum or coir dust which has been soaked in water and squeezed by hand. The wet sphagnum or coir dust is covered with a sheet of plastic tied at each end with twine or with a sheet of aluminium foil tightly twisted closed at both ends.

Callus and roots will form at the upper end of the girdled area in six to eight weeks. When the roots are well formed, the marcots are cut from the tree and either planted direct in the field or first planted in containers of soil and later set in the field.

Ground Layering

In this method, a branch of plant, which is near the ground, is chosen and a ring of bark about 2.5 cm is diameter is removed just below the bud. This branch is then bended and buried in soil when still attached to the mother plant. The soil is regularly watered to keep it moist. Within a few weeks, the roots are formed and new plant is separated from the mother plant. Separation should be done in such a way that the roots formed also go with the detached plant. These new plants should preferably be planted in pots or nursery rows for development of better root system and shoot system before planting in the fields. For example baramasi lemon.

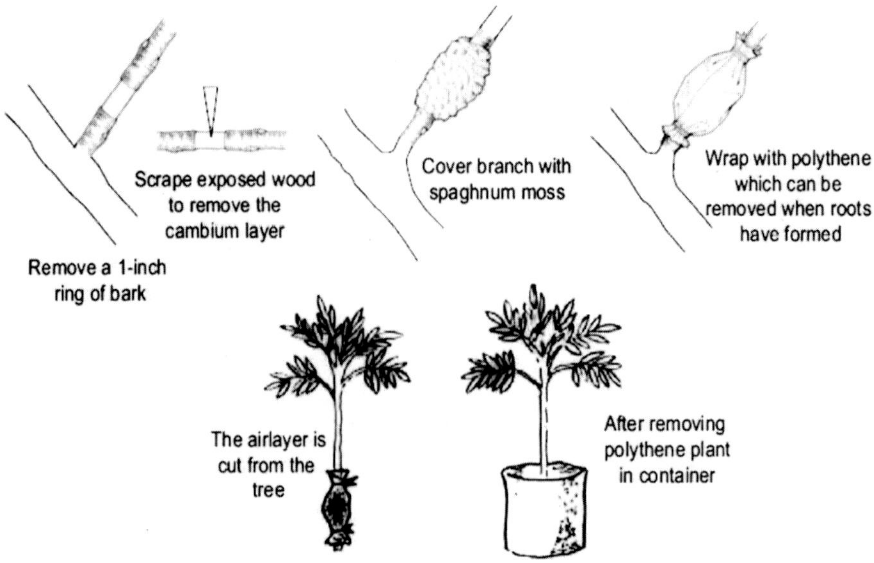

Fig. 1 Process of Air-layering

Mound Layering

In this method, plant is headed back either in February or in July. The new shoots come out during April and September, from ground level. A ring of bark is removed from these shoots and they are covered with moist soil (see figure 2). The rooted stools of April stooling are separated during rainy season and those of August are removed in the following spring. These stools, after separating from the parent plant are planted in the nursery fields. This method is also known as stool layering and is used for propagation of guava and apple rootstocks.

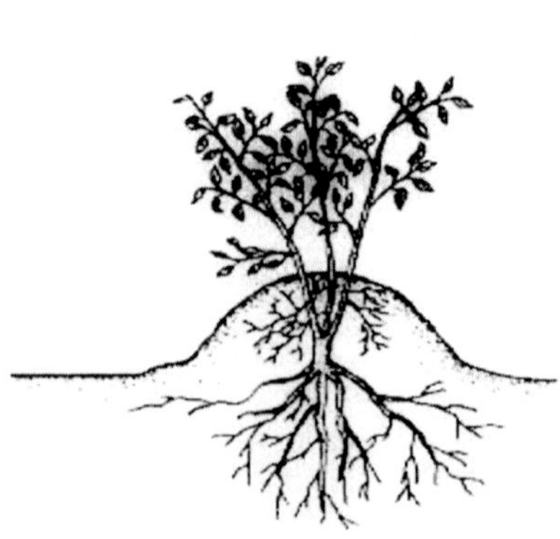

Fig. 2 Mound Layering

Grafting and Budding

Grafting is the joining of parts of plants together in such a way that they unite and continue to grow as a single plant. The part of the plant that becomes the upper portion or top of the new plant is called the scion and the part which becomes the lower portion that includes the root is called the stock or root stock.

All methods of joining plants are called grafting but when the scion is a small piece of bark or wood containing a single bud, this form of grafting is called budding. Many different types of grafts are possible including veneer, cleft, arching, bottle, tongue, whip, and root grafting.

Grafting requires experience and practice and the following six requirements to improve the performance of growers:

1. The stock and scion must be compatible or they will not unite. Graft only closely related species or plant families.
2. Cambial regions of scions and stock must be in intimate contact (see Figure 3). Cut surfaces should be held tightly for proper healing and flow of water and nutrients.

– – – – – – – – – – Cambial Zone

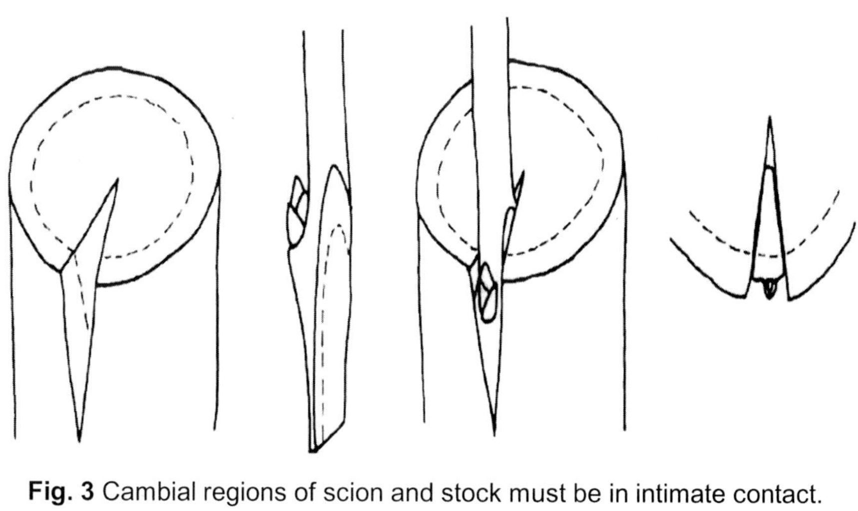

Fig. 3 Cambial regions of scion and stock must be in intimate contact.

3. Grafting can be done at any time of the year but March through August is considered ideal if the stock and scion are at the right size and growth stage. Some months like May and October are

however too wet and the scions show a tendency to rot unless special precautions are taken.

4. After grafting, all cut surfaces must be protected from dessication or drying out. This can be done by covering the graft area with wax or tape or some moist material like sphagnum moss.
5. Proper care must be given to the graft until it unites. Shoots from the stock must be removed as they can choke out the scion. Shoots from the scion can grow so vigorously that they break the scion off unless staked or tied.
6. The grafting knife should always be kept razor sharp during grafting operations.

Veneer Grafting

For the veneer graft, the root stock should be in an actively growing condition and scions should be clean and free of disease and insects. Scions in which the terminal bud is just beginning to swell will give best results. Root stocks 6 to 12 months old with stem diameters of 6 to 25 mm (1/4 to 1 inch) are used in commercial nurseries for side veneer grafting mangoes.

Select a site for the graft where the root stock is straight and 30 cm (12 inches) or less from ground level (see Figure 4). The cut is made tangentially into and through the bark and just into the wood for a length of 5 to 8 cm (2 to 3 inches) cutting down in one motion parallel to the wood.

A notch is formed with a separate horizontal cut at the lower end of the first cut in which the base of the scion will be placed. A slanting cut on one side of the scion is made, starting just below the terminal bud and continuing all the way to the base in one downward motion. A small wedge is made at the base on the opposite side of the scion to fit into the notch of the stock.

The scion is then placed in position so that the cambium of each piece joins as much as possible. A thick vinyl budding strip is securely wrapped around the stock and scion, starting from the bottom and working upwards over the terminal with enough overlap in each turn to keep out rain. The wrap is fastened by firmly looping the free end under the last turn. If the graft union is successful, new growth on the scion will begin in 2 or 3 weeks. The wrapping should be removed from the terminal as soon as growth starts and refastened just below the terminal. The vinyl strip may be removed after the first flush of growth has occurred and the rootstock partially cut back. The root stock may be cut off completely after the second flush. It should be trimmed in such a way as to leave no part of the rootstock above the union and buds and new growth on the rootstock should be removed as they appear.

Cleft or "V" Grafting

In Cleft or "V" grafting or wedge grafting (See Figure 5) the top of the root stock is cut off square and a vertical cut made in the centre using a heavy knife or special grafting tool. The scion is made by cutting a long, gradually tapering wedge. Holding the split open with a screw driver or similar tool, one or two scions are inserted into the split so that the cambial layers are lined up with each other at least on one side or both sides in the case of small plants of equal size. Wrapping and subsequent care is the same as for veneer graft but the root stock does not require cutting back and hence wound healing is faster. Make sure the scion and buds are not upside down.

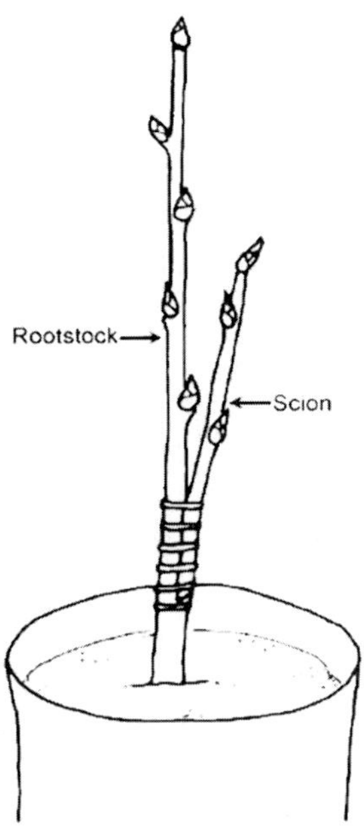

Fig. 4 Completed side veneer graft.

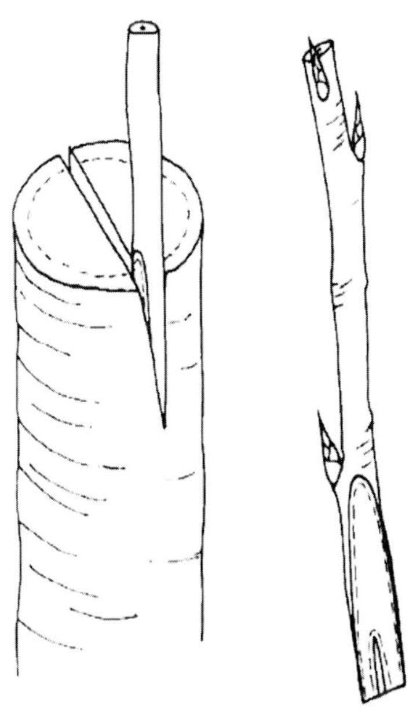

Fig. 5 Cleft or V Graft.

Tongue Grafting

This method is commonly used when the stock and scion are of equal diameter. First, a long, smooth, slanting cut of about 4 to 5 cm long is made (Figure 6) on the rootstock. Another downward cut is given starting approximately 1/3rd from the top and about centimeter in length. Similar cuts are made in the scion wood exactly matching the cut given in the rootstock. The scion having 2 to 3 buds is then tightly fitted with the rootstock taking care that the cambium layer of at least one side of the stock and scion unites together. This is then wrapped with polythene strip.

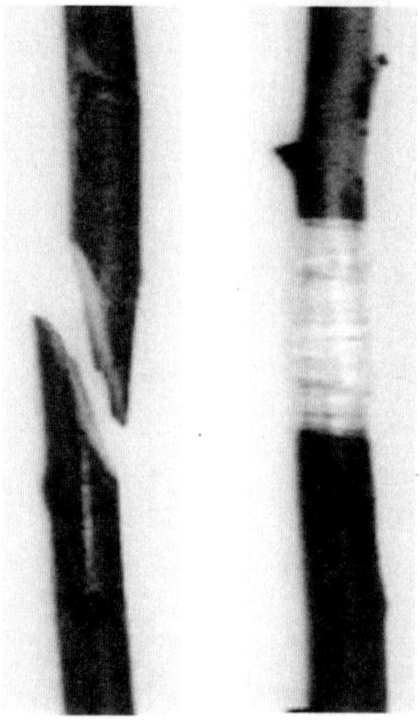

Fig. 6 Tongue grafting (a. Making tongue in stock & scion; b. grafting and tying with polythene strip).

Approach Grafting

This method of grafting is termed approach grafting, as the rootstock is approached (Figure 7) to the scion, while it is still attached to the mother plant. Alternatively the mother plants are trained to be low headed and the stock is sown under their canopy. Last week of July or the first week of August is the best period for approach grafting. In this method the diameter of rootstock and scion should be approximately the same. A slice of bark along with a thin piece of wood about 4 cm long is removed from matching portions of both the stock and the scion. They are then brought together making sure that their cambium layers make contact at least on one side. These grafts are then tied firmly with polythene strip or any other tying material. The stock and scion plants are watered regularly to hasten the union. The union is complete in about 2 to 3 months.

Fig. 7 Approach grafting

A cut is then given to the scion shoot about half way through its thickness. If the shoot does not show any sign of wilting for a week or so it is completely detached from the mother plant. In case the scion starts wilting it shows that the union is not complete. In such cases the scions are detached from the mother plants after some days when the union is complete. This method is also called inarching. The method is commonly followed in mango.

Side Grafting

A three sided rectangular cut about 4 x 1.25 cm is made on the rootstock at a height of about 15-20 cm from the ground level and the bark of the demarcated portion is lifted away (Figure 8) from the rootstock. A matching cut is also made on the base of the scion to expose cambium. The scion should be prepared well before the actual grafting is done. The healthy scion shoots from the last mature flush are selected for this purpose. The selected scion shoots should have plump terminal buds. After the selection of the scion shoots, remove the leaf blades, leaving petioles intact. In about 7 to 10 days the petioles shall drop and terminal buds become swollen. At this stage the scion stick should be detached from the mother tree and grafted on the stock. The prepared scion is inserted under the bark flap of the rootstock so that the exposed cambia of the two components are in close contact with each other. The bark flap of the rootstock is resorted in its position. The graft union is then tied firmly with polythene strip.

After the completion of the grafting operation, a part of the top of rootstock is removed to encourage growth of the scion. When the scion has sprouted and its leaves turned green, the root stock portion above the graft union should be cut away. Side grafting can be carried out successfully from March to October; but success during the May and October is rather low. This method of propagation is commonly used in mango.

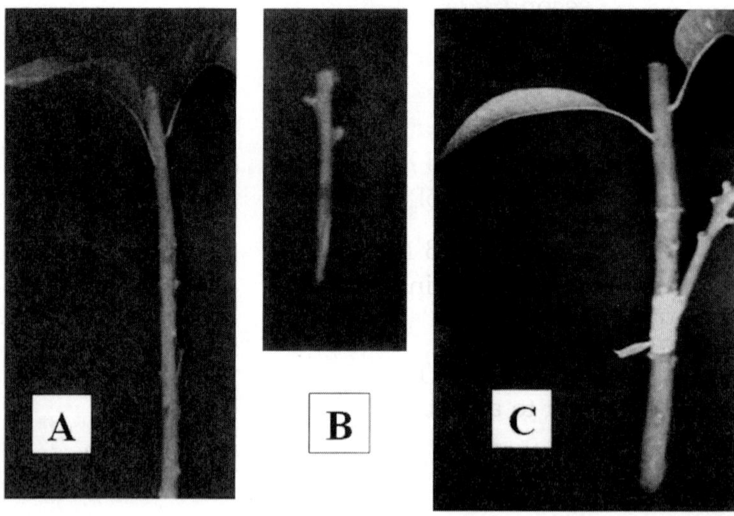

Fig. 8 Side grafting (A. preparation of rootstock: B. preparation of scion: C. grafting & securing the graft with polythene).

Budding

Budding is a method in which only one bud is inserted in the rootstock. This method is very easy and fast. This method saves budwood as compared to grafting. As soon as the bark starts slipping both on the stock and scion, this is considered to be the optimum time for budding. This shows that the cambium, which is the tissue responsible for union, is active. This method is generally employed during spring and rainy season. The common methods of budding are T-budding, patch budding, and chip budding.

Shield or Inverted T-budding

Budding differs from grafting in that only a single lateral bud is used instead of a portion of a stem with several lateral buds as well as a terminal bud. The principal advantage in budding is that one terminal scion will furnish five or more buds for as many trees.

The shield or inverted T-budding method involves the use of a bud that is shield shaped and an inverted T-cut into the stock with the following steps (see Figure 9):

1. A vertical cut 4 to 8 cm (1-1/2 to 3 inches) long is made on the stock and a second horizontal cut is made at the bottom of the vertical cut.
2. The bud is prepared by cutting into the scion 13 mm (1/2 inch) or more above the bud and cutting downward, going under the bud and coming out well below it, leaving a long handle on the lower part which is cut off after the bud is inserted.
3. The shield can be 6 to 13 mm (1/4 to1/2 inches) or more wide and 4 to 8 cm (1-1/2 to 3 inches) long with the bud being located in the centre.
4. The bark of the inverted T-cut is raised at the corners and along the vertical split to admit the shield.
5. The bud is forced into the cut and under the edges of the bark, being careful not to split the bark.
6. After insertion, the protruding handle of the shield is cut off with a horizontal cut at the bottom so that the shield can fit and slide completely into the cut made into the stock.
7. Buds are wrapped in 13 mm wide vinyl strips covering the bud completely.
8. Examine buds in two or three weeks and rewrap leaving the bud exposed if considered necessary.
9. Cut back top of root-stock gradually when bud begins to grow as well as any new growth below the graft.

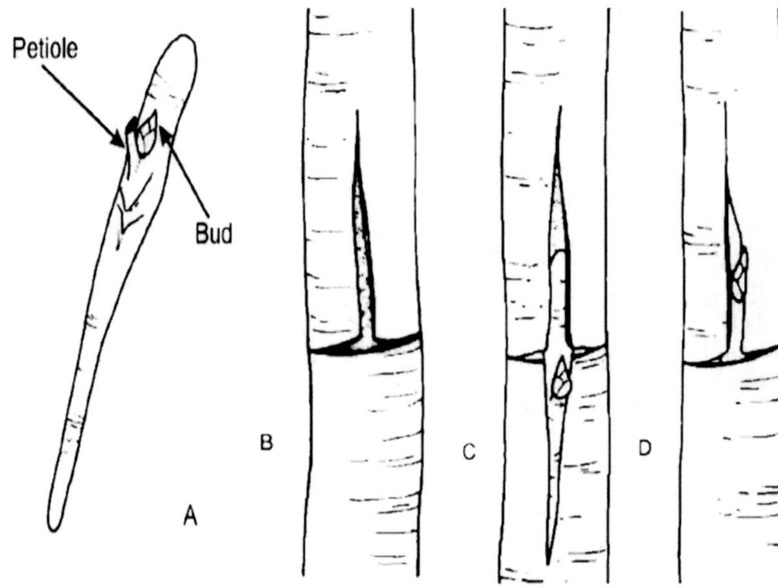

Fig. 9 Inverted T- or shield budding. A. The prepared scion. B. The vertical and horizontal cuts to form the inverted 'T' on the root stock. C. Insert the bud shield snugly up between the two flaps of bark on the root stock. D. The 'tail' of the shield is cut flush to the horizontal cut.

Hints in collecting Budwood

- The best budwood comes from the second flush of growth from end of branch
- Budwood should be rounded with plump gray-green buds; trim leaves to very small stubs as budwood is cut from tree
- Some buds can be difficult to grow, and girdling the particular branch and removing all but three or four leaves two to three weeks before the buds are needed can help pre-condition the bud and hasten growth.

Patch Budding

This type of budding is quite successful in guava and it gives 60 to 70 percent success during May and June. Freshly cut angular bud-wood from current season's growth should be used as scion. A rectangular or square patch or piece of bark about 1.0-1.5 cm broad and 2.5 cm long is removed from the rootstock (Figure 10) at about 15 to 20 cm from ground level. A similar patch with a bud on it is removed from the bud stick taking care not to split the bark beneath the bud. This patch is then transferred to rootstock and fixed smoothly at its new position and tied immediately with polythene strip. To have better success, a patch having two buds is used as scion instead of a single bud. This method is termed as improved patch budding method.

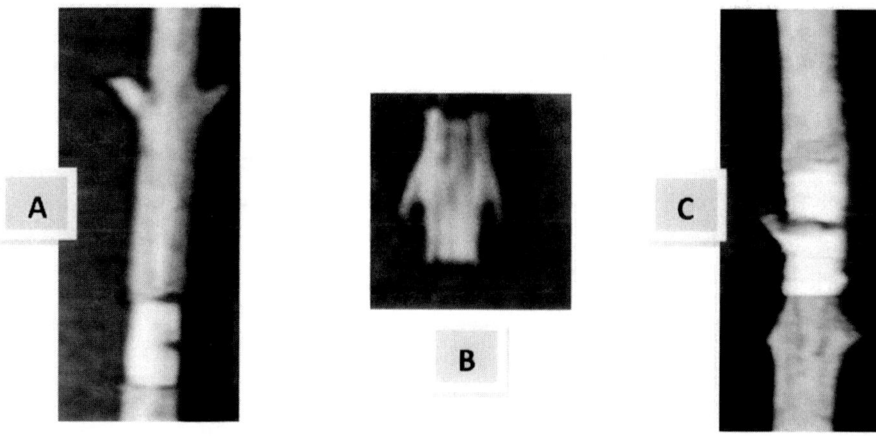

Fig. 10 patch budding (A. Removal of patch from rootstock; B. taking similar size patch from scion and C. fixing).

Chip Budding

This method is usually employed when the stock and scion are still dormant, that is just before the start of new growth. In this method, one about 2.5 cm long slanting cut is given into the stock followed by another cut at lower end of this first cut, in such a way that a chip of bark is removed from the stock. The bud from the scion wood is removed in the same way so that it matches the cuts given in the rootstock. This chip with a bud on it is fitted smoothly into the cut made in the rootstock taking care that the cambium layers of the stock and scion unite at least on one side. The bud is then tied and wrapped with polythene strip, to prevent drying up of the bud.

Propagation Through Specialized Organs

A. **Runners:** A runner is a specialized stem that develops from the axil of a leaf at the crown of a plant. It grows horizontally along the ground and forms a new plant at one of the nodes. The runner production is favoured by long day and high temperature. The daughter plants are separated and used as new planting material.

B. **Suckers:** A shoot arising on an old stem or underground part of the stem is known as suckers. In other words, a sucker is a shoot, which arises on a plant below the ground. These shoots, when separated from the mother plant and transplanted produce adventitious roots. The capacity of a plant to form suckers varies from plant to plant, variety to variety and is even climate dependent. The sucker formation is common in fruit plants like pear and banana.

C. **Slips:** Slips are shoots just arising below the crown but above the ground. Pineapple is commercially propagated through this method of propagation.

Special Techniques of Propagation

Apart from conventional methods of propagation such as sexual (seed) and asexual methods (Vegetative part), there are special techniques of propagation, which have played an important role in enhancing efficiency plant production. One of such methods is *in vitro*, also commonly referred as micropropagation.

Micropropagation

Sometimes referred to as tissue culture, micropropagation is a specialised method of mass production of seedlings in which very small pieces of plant tissue are regenerated in an artificial medium. Originally developed for ornamental plants, the technology has been extended to include forestry and fruit trees. Because of the high initial capital investment cost in laboratory and equipment and high salary and management costs, it is more economic to purchase such seedlings than to produce them.

The starting point for all tissue cultures is plant tissue or explant. This can be initiated from any part of a plant - root, stem, petiole, leaf or flower although the success of any one of these varies between species.

Pieces of plant tissue will slowly divide and grow into a colourless mass of cells if they are kept under special condtions which include:

- The presence of a high concentration of auxin and other growth regulators in the growth media.

- A growth medium containing organic and inorganic compounds designed to both sustain the plant cells, encourage cell division and control the development of either pale brown lumps called callus or as individual or small clusters of cells in a liquid medium called a suspension culture. These cells can be maintained indefinitely provided they are sub-cultured regularly into fresh growth medium.

For tissue culture to be successful, the parent must be healthy and free from obvious signs of decay or disease. The plants themselves must be actively growing and not about to enter a period of dormancy. It is essential the surface of the explant is sterilised to remove all forms of contamination. Since cell division is slow compared with the growth of bacteria and fungi, even minor contaminants will easily overgrow and destroy the plant tissue culture.

The exact conditions required to initiate and sustain plant cells in culture, or to regenerate plants from cultured cells, are different for each plant species. Each variety of a species will often have a particular set of cultural requirements. Despite all the knowledge that has been obtained about plant tissue culture, these conditions have to be identified for each variety through experimentation.

Advantages: Micro propagation has the following advantages over conventional propagation techniques:

- Year around production of plants irrespective of seasonal constraints.
- Small space is required to maintain and multiply large number of plants.
- Small tissue is required as an explant, hence saves the scion wood to a great extent.
- Speedy international exchange of germplasm, requiring minimum quarantine checks is possible.
- Micropropagated plants exhibit vigorous growth, and higher yields.
- Micropropagated plants are usually free from viruses.
- Micro propagation is highly beneficial in dioecious fruit plant species (date palm and papaya), where large-scale production of female plants is possible.

- It helps in reducing the breeding cycle.
- Production of homozygous plants is possible under *in vitro* conditions.
- It is highly beneficial in plants in which vegetative propagation is not possible or the propagation rate is very slow (papaya and date palm).
- *In vitro* systems have the potential for long-term transportation or shipment of propagation material.

Disadvantages

- Expensive and sophisticated facilities, trained personnel and specialized techniques are required.
- High cost of production results from expensive facilities and high labour inputs.
- Contamination or insect infestation can cause high losses in a short time.
- Higher level of somatic variation.
- Poor establishment of the plantlets in the field.

Stages of micropropagation

Micro propagation is an integrated process in which cells, tissues or organs of the selected plants are isolated, surface sterilized and incubated in growth promoting, sterile medium and environment to produce a large number of plants. The different stages are:

Stage 0: Selection of mother plant for explants isolation: The mother plant from which explants has to be excised should be

a) A certified and true to type representative of the desired species and or cultivar.

b) Healthy and free from insect pest and disease.

c) Should be quite vigorous.

Stage 1: Explant establishment in culture medium: During this stage the explant is cultured in a suitable culture medium, preferably agar based media for tissue activation and multiplication.

Stage 2: Proliferation and multiplication: In this stage, repeated sub-culturing is done to encourage more proliferation, which largely depends upon the combination of growth regulators. The duration of this stage is unlimited and largely depends on the choice of propagator.

Stage 3: Plant establishment and rooting: In this stage the selected plants are forced for root formation, which can be achieved by media modification and modifying the concentration of growth regulators. The concentration of cytokinins and sugars are reduced and concentration of auxins and light intensity in the laboratory is increased to start with photosynthesis and other physiological activities.

Stage 4. Acclimatization or hardening: The plantlets developed in the culture tubes are acclimatized to specific environment having a high humidity, a low light level and a constant temperature. Besides, the roots developed *in vitro* are hairless and hence delicate, requiring care during transfer from culture medium. To have better survival rate, the plantlets may be transferred to container kept in mist chambers where relative humidity is maintained at higher order. Once new growth is seen, the plants may be slowly transferred to outside by exposing to increased light intensity in stages.

Once plantlets are well rooted, they must be acclimatized to the green house environment. *In vitro* rooted plants are removed from the culture vessel and the agar is washed away completely to remove a potential source of contamination. Plantlets are transplanted into a standard pasteurized rooting or soil mixture in small pots or cells in a more or less conventional manner. Initially, microplants should be protected from desiccation in a shaded, high humidity tent or under mist or fog. Several days may be required for new functional roots to form.

Plantlets should be gradually exposed to a lower relative humidity and a higher light intensity. Any dormancy or resting condition that develops may need to be overcome as part of the establishment process. These conditions help the plantlets in getting acclimatized to the natural conditions, which would help them in getting established in the field easily.

Culture Techniques

Various culture techniques such as (i) meristem culture (ii) callus culture (iii) shoot bud regeneration (iv) somatic embryogenesis (v) ovule culture (vi) embryo culture (vii) anther culture and (viii) protoplast culture are employed in micropropagation.

1. **Meristem culture:** Meristem culture involves culture of both shoot-tip and axillary-bud. The use of small shoot-tips comprising of the apical dome with one or two leaf primordia (0.1-0.5 mm) is the basis for the technique known as meristem-tip-culture, pioneered by Morel in the 1950s. Meristem tip culture is now being routinely used, mainly in horticultural crops, for the elimination of virus from infected material. Virus apparently either does not easily invade or rapidly multiply in the young meristematic tissue. A simple nutrient medium consisting only of salts, sucrose and vitamins is used in order to minimize the formation of callus. Gibberellic acid is often needed to promote adequate growth and NAA may be required to stimulate root formation.

2. **Callus culture**: *A* piece of sterile plant tissue with living cells is transferred to a culture medium to induce callus proliferation. Sub-culturing is then done onto a medium with or without altered growth regulator concentrations, ultimately resulting in the induction of adventitious organs or embryos. In the last stage, regenerated plants are removed from *in vitro* culture and slowly exposed to outer environment so that the plants can be fully autotrophic.

3. **Cell culture**: The cells are maintained in suspension cultures so as to produce free cells and are then sub-cultured to regenerate complete plant from single cells. This technique is now useful to induce variability in plant cells and to select desirable cell variants and regenerate complete plants from these variants.

4. **Embryo culture:** It involves aseptic excision of the embryo and its transfer to a suitable medium for development under optimum culture conditions. After the embryo has grown into a plantlet *in vitro,* it is transferred to sterile soil or vermiculite and grown to maturity in a green house. This technique is useful in the production of intetspecific and intergeneric hybrids which could not be otherwise accomplished and also in overcoming embryo abortion.

5. **Protoplast culture**: From different sources, protoplasts (the plant without any rigid cellulose wall but with plasma membrane only

allowed to fuse to form a somatic hybrid) are cultured in suitable media to regenerate the cell wall and are again cultured in suitable medium for differentiation and morphogenesis.

6. **Anther culture:** The culture of anthers is of considerable value to breeders as it is possible to produce haploid plants which reveal recessive alleles. These haploid plants can be used for the production of homozygous diploids, thus avoiding generations of inbreeding. Added benefits, such as small flowers and prolonged flowering time, might be ensured from the use of haploid plants as they are usually smaller than their diploid counterparts and being sterile there will be no pollination induced senescence. Anther culture has been used in *Pelargonium* spp. to eliminate virus, in *Lilium* spp. to produce haploid plants and in *Gerbera* to obtain different flower colour.

7. **Somatic embryogenesis:** The greatest potential for clonal multiplication is through somatic embryogenesis, where technically a single isolated cell can produce first an embryo, then a complete plant. Somatic embryogenesis and plantlet regeneration has been reported in various species of horticultural plants by using mid-rib, leaf and stem callus on modified MS basal medium supplemented with 1.0 - 2.0 mg/l 2,4-D and 0.25-0.50 mg/l BA or kinetin.

Culture medium

Success in the technology and application of tissue culture methods depends on the selection of proper culture medium, which meets the nutritional requirements of cultured cells and tissues. The basic components of all nutrient media are inorganic salts, carbohydrates, vitamins, growth regulators, agar (for solid medium) and water. Other components including organic nitrogenous compounds, organic acids and complex substances can be important but are optional. The inorganic nutrients required in macro-amounts are N,P, K, Ca, S and Mg. The optimum concentration of each nutrient for achieving maximum growth rates varies considerably. The essential nutrients required in micro-molar concentration include Fe, Mn, Zn, B, Cu and Mo. The concentration of inorganic nitrogen in the culture medium varies from 25 to 50 mM. Nitrate is commonly used in the range from 25-40 mM and the amount of ammonium varies between 2 and 20 mM. Potassium is supplied in the range of 20 mM or higher and the optimum concentrations of P, Mg, Ca and S for most tissues are 1-3 mm.

The standard carbon source is sucrose or glucose at a concentration of 2-3%. Other carbohydrates which have been tested include fructose, lactose, maltose, galactose and starch but these compounds are generally much inferior to sucrose or glucose as a carbon source. Most media contain myo-inositol. There is no absolute requirement of myo-inositol, but its inclusion (100 mg/l) generally improves tissue growth. MS medium is the most popularly used nutrient medium in tissue culture.

The nutrient media are supplied with certain vitamins. There is an absolute requirement (0.1 - 0.4 mg/1) for thiamine. Growth is also improved by the addition of nicotinic acid and pyridoxine; sometimes biotin, vitamin B12, folic aci d, choline chloride, riboflavin and ascorbic acid are also added to the culture medium, depending on the type of tissue to be cultured. The pH of the culture medium is an important factor for maintenance of growth of cultured plant tissues. In general, plant tissues require an acidic pH, and an initial pH of 5.5 - 5.8 is optimum. The physical form of the culture medium is also an important factor determining the success or failure of tissue culture. Both liquid and solid form of culture medium are useful for initiation and proliferation of callus tissue and initial organogenesis, but for further growth of organs solid medium is essential. When culturing in a liquid medium, gentle agitation of the liquid is necessary for aeration of culture medium, rotating apparatus is used for this purpose.

Important Term and Terminology

Selection of Elite Mother Trees

- It should be of known identity
- It should have production potential
- It should have commercial acceptance
- It should be free from pests and pathogens

Selection of Scion Wood, Rootstock and Interstock: The selection of scion wood, rootstock and interstock (the stock between scion and rootstock) should be done carefully. The selection criteria vary in each case.

Selection of Scion Wood

- The scion should be from mature shoot i.e. at least one year old.
- A scion wood of diameter 0.6-1.2 cm is satisfactory for better bud wood.
- The scion shoot should have healthy, well-developed round and plump buds.
- Scion should be selected from elite trees known for quality production of fruits.
- Scion wood should be free from any bacterial, fungal and viral diseases.
- The scion should be dormant, while selected for grafting on rootstock.
- The best scion wood can be obtained from the central portion or from the basal portion of shoot. The terminal sections, which are generally succulent, pithy and low in store of carbohydrate (CHO), should be discarded.

Collection and Storage of Bud Wood: The best quality scion wood usually comes from shoots grown in the previous season. Scions should be severed with sharp, clean shears or knives and placed immediately in moistened plastic bags. It is good practice during the harvesting of scions and the making of grafts to clean the cutting tools regularly. Flaming or immersing them in a sterilizing solution may do this. Isopropyl alcohol also works well as a sterilant, although it evaporates quite readily. An alternative sterilizing solution may be prepared by mixing one part household bleach with nine parts water (by volume) However, this bleach solution can be highly corrosive to certain metals. For best results, harvest only as much scion wood as can be used for grafting during the same day. If large quantities of scion wood must to be harvested at one time, follow these steps:

- Cut all scions to a uniform length, keep their basal ends together, and tie them in bundles of known quantity (for example, 50 scions per bundle).
- Label them, recording the cultivar, date of harvest, and location of stock plant.
- Wrap the base of the bundles in moistened burlap or sphagnum, place them in polyethylene or waterproof paper bags, and seal the bags.

- Store the bundles for short periods, if necessary, either iced down in insulated coolers or in a commercial storage unit at 32° to 34°F.
- Never store scions in refrigerated units where fruits or vegetables are currently kept or have been stored recently.
- Keep the scions free from freezing during storage.

Selection of Rootstock

1. Rootstock should have a proper vigour and growth habits.
2. Rootstock should be resistant to soil born diseases and other pests.
3. Rootstock should be tolerant/ resistant to toxic salts like Na, Mg, and Ca etc.
4. It should have wide range of edaphic adaptability (all soil and climatic factors).
5. Should have wide range of graft compatibility.
6. Should be easy to propagate.
7. It should not go under any mutation.
8. Its age should be one to one and half years and not more than 2years.
9. Its diameter should be greater than 1 cm (pencil thickness).

Selection of Interstock

Interstock is a piece of stem inserted by means of two graft unions between the stock and scion. It is used to

1. Avoid incompatibility between the stock and scion.
2. Make use of winter hardy trunk.
3. Take advantage of dwarfing rootstock. e.g. Apple: to restrict tree size and improve rooting.

Bud Wood Certification: The commercialization of different fruit plants and with the urge to produce superior quality cultivars, there is uncontrolled importation of varieties within the country and across the countries. Such movement of budwood across the globe results in spread of viral diseases from one region to the other, which needs to be controlled. Bud wood certification has a prominent role in this direction, provided if done in each and every case of germplasm movements within and across the countries.

Bud Wood Packaging: The bud wood should not be allowed to dry out and be kept in polythene bags containing damp sawdust/ wrapped in moist pieces of gunny bags/polythene bags Although, essential requirements are to keep the wood moist and viable, but not so wet as to rot, and maintaining a cool enough temperature to prevent premature bud swelling and protecting against freeze damage. For temperate fruits store scions in tightly sealed plastic bags with some moisture in refrigeration until ready to use.

Bud Wood Labeling: Clearly labeled outside packaging where there is no regulatory guidance, the label may include:

1. Type of material
2. Plant species with cultivar
3. Date of harvest
4. Permit Number (where applicable)
5. Permit Number for import and/or Phytosanitary Certification (where applicable).
6. Amount of material shipped
7. Location of the stock plant
8. Genetically modified or not

❑❑❑

8

Plant Protection in Nursery

Introduction

The need to produce more food to feed the increasing population will continue. The systems and technologies that are being used now may have to be modified and new technologies adopted in order to ensure that this goal can be achieved. Pests are a major bottleneck in attaining this goal. Pesticides will continue to play an important role in protecting the crops from pests in the foreseeable future, as there are no practical alternatives at the moment. However, of late there have been many encouraging developments that give hope for the future.

Pest Management in Nursery

If you ever visit a garden or crop field you will come across cuttings on leaf margins, holes in leaves, chewed and damaged parts of plants, spots on leaves etc. They are due to the attack of various organisms which are known as pests. A pest is any organism, animal, plant or microorganism that causes damage to the plants, animals or human beings. The word pest is derived from the Greek word 'pestis' which means to annoy, to disturb or to destroy.

A major injury to nursery stock is also caused by various groups of insects. These insect pests have been divided into three categories viz., major nursery pests (white grubs, cutworms, termites and crickets), minor nursery pests (defoliators, sapsuckers, grasshoppers) and non-insect pests (nematodes and vertebrate pests). Generally the damage caused by the insects may be controlled by maintaining better sanitation of the nursery area, adoption of suitable cultural practices and need based application of chemical and biological pesticides.

White Grubs: The adult white grubs feed on leaves and larval stage of the grub (during monsoon months feed on roots. It is a major pest in Teak, Mango, Sal and leguminous seedling. Deep ploughing, soil solarisation, poisoning and using light traps are some control measures against white grub attack. Application of 200 g phorate or 50 ml of chloropyriphos mixed in 50 ml water may be used to spray for one bed. Foliar spray of host trees available in the nursery vicinity with 0.03% quinalphos can also helpful in controlling the adult population.

Cutworms: It damages the young seedlings soon after germination and is also a feeder of young leaves. Seedlings of Pine, Cedar, mango, sapota and *Casuarina* species are the most preferred by cutworms. Nursery site flooding and collection of cutworm after heavy rains are some preventive measures to avoid cutworm damages. Dusting of seed bed with a mixture of quicklime and ash or 1.5% quinalphos will control the insect.

Termite: They cause damage to seedlings either by primary attack (tap root destroy), secondary attack (follow up attack after draught, pathogens, etc.) or complementary attack and damage the seedlings which make it weak and subsequently it is susceptible for other pathogen and pest attacks. The termite attack can be controlled by keeping the nursery cleared of wood debris, using well decomposed FYM and application of termiticides such as chlorpyriphos.

Crickets: The nymphs and adult stage cricket come out at night and cut off all the seedlings, low branches and drag the piece to their tunnels for feeding the young crickets. Ficus, *Casuarina*, Eucalyptus, Sisham, teak, rubber and mango seedlings are commonly affected by crickets. Deep ploughing during nursery site preparation, application of 200 g phorate or fenitrothinon 5% dust per bed can control the pest.

Minor and Non-insect Pests: Defoliators (beetles, weevils and caterpillers), grasshoppers and sapsuckers (green leaf hopper, white flies, thrips) are the minor pests. They can be controlled by the application of 100 g dose per bed of phorate 10%, or spray of formulation of any systemic insecticide *e.g.* dimethoate 30 EC.

Nematodes, rat, squirrel, hare, deer, mite and birds are some important non-insect pests. Poison bating by rodenticide such as Zinc phosphide, proper fencing and manual scaring are the best methods to reduce damage by them. Other than the disease and pest damages, the natural events like frost, chilling, drought, fire and non-availability of nutrients also cause stunted growth/death of seedlings.

Methods of Pest Control

1. **Cultural Method:** It refers to manipulation of farm practices to check the pests. Some of the important cultural methods are: Crop rotations, Tillage methods (deep summer ploughing), High seed rate, Water management, Manipulation of date of sowing, and Trap cropping.
2. **Physical Method:** These methods involve modification of physical factors in the environment to minimize or prevent pest problem. Various physical methods are: Temperature manipulation, Moisture manipulation, Light manipulation, and Use of sound.
3. **Mechanical Method:** This refers to the use of mechanical implements and devices for removal and destruction of pests. Some of them are, Screens, traps, nets and suction devices, Hooking devices with iron rod in the hole bore by the insect, Banding with grease or polythene sheets on stem, Covering of seedling with net; and trenching and water barrier-ant pans.
4. **Legal or Regulatory Method:** This refers, to the legal restrictions proposed by the Central and State Governments to check the spread of pests. The regulations such as Inspection and quarantine and Destructive Insect Pests Act can be enlisted under this heading.
5. **Resistant Varieties:** Use of resistant varieties help in avoiding or tolerating or recovering from pest attack. Resistant varieties have been identified against various pests in a number of crops.
6. **Biological Method:** This method refers to the use of natural enemies of pests viz. parasites, predators and microbes or pathogens (bacteria, virus, nematodes, fungi, protozoa etc.) so as to suppress the pest species. Biological control program can be carried out in the following ways: conservation and encouragement of indigenous natural enemies, importation of exotic natural enemies and mass rearing and releases of parasites/predators and microbes.

7. **Chemical Control:** The use of chemicals for the control of pests is known as chemical control. Pesticides are the chemicals used to kill or repel or attract or sterilize pests. Pesticides are without any doubt an effective means of killing pests quickly and on demand. No other control method provides users with an immediate and visibly effective means of pest control. Over the years, the indiscriminate use of pesticides was resulted in a number of serious detrimental effects on the environment.

8. **Use of Botanical Pesticides:** These include the use plant products with a potential to control pests. Many plant products (leaf extracts, oils and cakes) have the property of inhibiting the development of pests and diseases. The plant extracts and oils are sprayed on the crops. Neem oil, neem cake and other neem based formulations have been found effective against pests.

Use of Pesticides in Pest Control

The development of effective, economical pesticides has had a profound effect on man's continual battle with pests. In many cases pesticides have been incorporated as tools in well planned pest control programmer without serious hazards to humans or to the environment. Application of pesticides must be done at proper time, at right rate by using suitable equipment. The pesticides are applied on seeds, foliage and other parts or in soil against different pests and diseases. The various methods of pesticides application are Seed treatment, foliar application, Soil application, Granular application, Seedling root dip, Fumigation, Baiting etc.

Table 1. Important nursery pest and their control measures

Pest	Damage	Management
Pseudostem Borer	The grubs bore into the stem and feed within the stem.	Application of Carbaryl WP (0.1%) also controls the pest population.
Beetle	They make holes in cotyledonary leaves. As a resultthe seedlings die in the youngerstage	Spraying Carbaryl (4g/liter of water) or Metacid (1ml/liter of water)
Aphids	Aphids damage the plants bysucking the	Spraying Imidadopod (0.03%)

	leaf sap in young-stage, cotyledonary leavescrinkle and in severe cases theplants withers off.	Dinethafe (0.2%)
Jassids	Both nymphs and adults suck thesap from the lower surface of theleaves. The infested leaf curl upward along the margins, which may turn yellowish and show, burnt up patches.	Spraying Malathion (0.1%) or Dichlorvos (0.05%)
Leaf Roller	Caterpillars roll leaves and feedon chlorophyll while remaininginside the folds. The foldedleaves wither and dry up.	Spraying of Carbaryl (0.1%) orMalathion (0.05%)
Red Spidermite	Different stages of mites arefound in colonies covered bywhite-silky webs on lowersurface of leaves. Nymphs and adults suck cell sap and whitepatches appear on leaves. Affected leaves become mottled, turn brown and fall.	Acaricides like Omite (0.05%) and Wettable Sulphur (0.3%) gives effective control of mites.
Mealy Bug	Nymphs and adults of mealybugs suck sap from the leaves,tender shoots and the fruits. Aheavy black sooty mould maydevelop on the honeydew likedroplets secreted by mealy bugs.	Spraying of insecticides like Dichlorvos (0.02%) or Chlorpyriphos (0.05%) with fish oil rosin soap wasfound to control the insect population.
Root-Knot Nematodes	The affected plants show thedevelopment of galls on theroots. The plants become stuntedand the leaves show chlorotic symptoms.	Treating the nursery beds with Phoret@ 5 g a.i./m2 or Neem Cake 1 kg/m2Select resistant varieties.
Cabbage Diamo -ndback Moth	Severely affected leaves arecompletely skletonised.	Spraying the crop with Malathion (0.1%) or Profenofos (0.25-0.5 kga.i./ha)

Thrips	Nymphs and black adults feed on tender leaves causing silvering, mottling and distortion of leaves.	Soil application of thimate twice at 15 days interval at 5 gm/bed and alsotake spray.
Leaf Folder	Larva fold leaves together and feed on chlorophyll. The affected leaves get skeletonized and dry.	Two or three sprays of Propenophos @ (0.1-0.2%)
Leaf Miner	Larvae attack tender leaves and feed in the epidermal layers of the leaf by making serpentine-mines in which air gets trappedand gives them silveryappearance.	Spraying the plants with Profenophos @ (0.1-0.2%)
Leaf Eating Caterpillar	Larvae feed on lower surface of leaves by scraping whilegreenish-brown mature larvaefeed voraciously during nightson these leaves.	Spraying of Profenophos @ (0.1-0.2%) or Carbaryl @ 0.1% or Chlorpyriphos @0.05%
Cutworms	The tender plants are found damped at ground level duringthe night Young larvae feed gregariously on foliage but latersegregate and enter into soil.	Soil application of Phorate (1kga.i./ha)
Whitefly	The damage by whitefly alsoleads to yellowing of leaves andstunted growth, in severe casesleading to shedding of leaves	Spraying Triazophos 40 EC (1.5 ml/L of water) + 1.0 ml of Dichlorvos 76 EC per litre of water.

Nursery Disease and their Management

The seedling stress symptoms like damping off, wilt, root rot, rust and powdery mildew are caused by pathogen infection and results in stunted growth of seedlings. These pathogens may be soil, seed or air borne in nature. Nurseries established in the recently cleaned land hardly invite parasitic organisms. Stunted growth of seedlings indicates the loss of soil fertility, excess watering and dumping of seedlings in shady areas.

As a preventive measure sterilization of nursery mixture, pre-treatment of seeds with fungicide such as Captan can control the disease. If the disease occurs, the casual pathogen may be identified by expression of symptoms and accordingly fungicide may be applied. Table 2 listed the important nursery diseases and their control measures.

Favorable Conditions for Diseases Infestation

Plant diseases are caused by various sources like micro-organisms, including fungi, bacteria, viruses, mycoplasmas, etc. or may be incited by physiological causes including high or low temperatures, lack or excess of soil moisture and aeration, deficiency or excess of plant nutrients, soil acidity or alkalinity, etc.

The causative agents of disease in green plants are in thousands and include almost every form of life. But primary agents of disease may also be inanimate. Thus nonliving (abiotic) agents of disease include mineral deficiencies and toxicities, air pollutants, biologically produced toxicants, improperly used pesticide chemicals, and other environmental factors as wind, water, temperature and sunlight.

Abiotic factors: Nonliving agents certainly qualify as primary agents of disease; they continuously irritate plant cells and tissues; they hamper the physiological processes of the plant; they evoke pathological responses that later show up as symptoms and characteristic of the different diseases. The abiotic agents of plant disease are termed noninfectious and the diseases they cause are termed noninfectious diseases.

Micro-organisms: The micro-organisms obtain their food either by breaking down the dead plant and animal remains (saprophytes) or by attacking living plants and animals (parasites). In order to obtain nutrients, the parasitic organisms excrete enzymes or toxins and kill the cells of the tissues of the host plant. These toxins kill or damage whole plant or a part of it, and cause considerable disturbance in its normal metabolic processes.

Parasites: The parasites are one of the major factors causing plant diseases. Parasites are those living organisms which colonize the living tissues of the host-plant. These diseases can be transmitted from plant to plant. These biotic agents are, therefore, infectious, and the diseases they cause are termed infectious diseases.

Ability to Produce an Inoculum

The parasitic pest must produce an inoculum, some structure that is adapted for transmission to a healthy plant and thus can establish a parasitic relationship with the host. For example, inocula for viruses are the viral particles (virions); for bacteria, the bacterial cells; for fungi, various kinds of spores or the hyphal threads of mold; for nematodes, eggs or second stage larvae.

Media for Transportation of Inoculum

The inoculum must be transported from its source to a part of a host plant that can be infected. This dispersal of inoculum to susceptible tissue is termed inoculation. Agents of inoculation may be insects (for most viruses, mycoplasma-like organisms, bacteria and fungi), wind and rain (for many fungi).

Wounds and Natural Openings

The parasite must enter the host plant to cause an infection. A parasite can enter the host through the following means, through wounds, through natural openings, or by growing directly through the unbroken protecting surface of the host. Viruses are injected into the plant through the proboscis of insects feeding on the juice of the host plant.Bacteria enter through wounds or natural openings like, stomates, hydathodes, and lenticels. But many fungi can penetrate plant parts by growing directly through plant surfaces, exerting enormous mechanical pressure and possibly softening host surfaces by enzymatic action.

Availability of Food

Availabitlity of food or nutrition within the body is essential for the parasite to grow within the host. This act of colonization is termed infection. The parasite damages the cytoplasmic memberanes of the host cells and makes the membranes freely permeable to solutes for nourishment of the parasite and parasitism. This is achieved by enzymatic attacks of the parasite upon carbohydrates, proteins and lipids inside the host cell. The breakdown products of such complex molecules would diffuse across the damaged hostcell membranes and be absorbed by the parasite in the form of sugars and amino acids.

Temperature Effects

The growing season of the parasite is affected by the temperatures of the surrounding. Many pathogen systems of cereal rusts in particular are affected by temperature. At times about half of the wheat cultivars showed differential resistance expression when tested against isolates of brown rust (leaf rust), either effective at 10°C and not at 25°C or vice versa.

Method of Disease Control in Nursery

A. Preventive Measures

Cultural Practices: Cultural practices usually influence the development of disease in plants by affecting the environment. Such practices are intended to make the atmospheric, edaphic, or biological surroundings favorable to the crop plant, unfavourable to its parasites. Cultural practices that lead to disease control have little effect on the climate of a region but can exert significant influence on the microclimate of the crop plants in a field. Three stages of parasite's life cycle namely, Survival between crops, production of inoculum for the primary cycle and inoculation can be controlled by following preventive measures.

1. **Survival between Crops**: Organisms that survive in the soil can often be controlled by crop rotations with unsusceptible species. Catch crop has been used to control certain nematodes and other soil-borne pathogens. Soil-borne plant pathogens can be controlled by biological methods. Plant parasites may be controlled by encouraging the growth of antagonistic organisms. This can be achieved by adoption of cultural practices such as green manuring and the use of appropriate soil additives. Soil-borne plant parasites may also be killed during their over-seasoning stages by such cultural practices as deep ploughing, flooding, and frequent cultivation and fallow. Plant diseases caused by organisms that survive as parasites within perennial hosts or within the seed of annual plants may be controlled therapeutically.
2. **Production of Inoculum for the Primary Cycle**: Environmental factors (particularly temperature, water, and organic and inorganic nutrients) significantly affect Inoculum production. Warm temperature usually breaks dormancy of over seasoning

structures; rain may leach growth inhibitors from the soil and permit germination of resting spores and special nutrients may stimulate the growth of seasoning structures that produce inoculum.

3. **Dispersal of Inoculum and Inoculation**: Cultural practices that exemplify avoidance are also effective in avoiding dissemination. Dispersal can also be avoided over a larger span with the help of plant quarantine. Plant quarantine is the legally enforced stoppage of plant pathogens at points of entry into political subdivisions.

4. **Sample Inspection**: One of the preventive measures to control the diseases is the use of sample inspection method. Laboratory evaluation of the representative sample drawn by the certification agency for the determination of germination, moisture content, weed seed content, admixture, purity, seed-borne pathogens.

Control Measures

Chemical Control: The pesticide chemicals that control plant diseases may be used in different ways, depending on the parasite to be controlled and the circumstances required for parasitic activities. Relatively insoluble protective fungicides are applied repeatedly to the green leaves of potato plants to safeguard them from penetration by the fungus of late blight. Also, systemic fungicidal chemicals may be used therapeutically. The oxathiin derivatives that kill the smut fungi that infect embryos is therapeutic. The Benomyl which has systemic action against powdery mildews and other leaf infecting fungi is also therapeutic in its action. Volatile fungicides are often useful as soil fumigating chemicals that have eradicative action.

The chemical control of plant diseases is classified in three categories: seed treatments, soil treatments and protective sprays or dusts.

1. **Seed Treatments:** Chemical treatments of seed may be effective in controlling plant pathogens in, on, and around planted seed. Seed treatment is therapeutic when it kills bacteria or fungi that infect embryos, cotyledons, or endosperms under the seed coat. Seed treatment is called eradicative when it kills spores of fungi that contaminate seed surfaces. Seed treatment is protective when it prevents penetration of soil-borne fungi into seedling stems.

Seed treatment is of two types; viz., physical and chemical. Physical treatments include hot-water treatment, solar-heat treatment, etc. Chemical treatments include the use of fungicides and bactericides.

These fungicides are applied to seed by different methods. In one method, the seed in small lots is treated in simple seed-treaters. The seed-dip method involves preparing fungicide suspension in water and then dipping the seed in it for a specified time.

2. **Soil Treatments**: Soil-borne plant pathogens greatly increase their populations as soils are cropped continuously, and finally reach such levels that contaminated soils are unfit for crop production.

   Chemical treatments of soil eradicate the plant pathogens and offer the opportunity for uninterrupted agricultural uses by rapid reclamation of infested soils. Preplanting chemical treatment of field soils for the control of nematode-induced diseases, fumigation of seedbed and greenhouse soils with methyl bromide, etc is commonly practiced to eradicate weeds, insects, and plant pathogens.

   Field applications of soil-treatment chemicals for fungus control are usually restricted to treatments of furrows. Formaldehyde or captan applied against sclerotia-producing fungi that cause seedling blights, stem rots, and root rots of many nursery seedlings. Other soil-treatment fungicides are vapam and "Vorlex". Soil treatments are made at the time of planting are most effective against parasitic attacks that come early in the growing season.

3. **Protective Sprays and Dust**: Protective fungicides prevent germination, growth and penetration. In order to use protective fungicides effectively, the farmer must not only select the right fungicide for the job but also apply it in the right amount, at the right times and in the right way. Too little fungicide fails to control disease; too much can prove toxic to the plants to be protected. The nurseryman and applicator, therefore, must always follow use instructions to the latter. Timing of applications is also critical.

Table 2. Important nursery diseases and their control measures

Symptoms	Affected seedling species	Control measures
Wilt, root rot, collar rot	Sisham, neem, Casuarina, Eucalyptus, tomato	Soil drenching with 0.1 % Carbendazim
Leaf spot	Eucalyptus, pomegranate	Spraying Dithane M 45 or Fytolan 0.2% or Copper oxychloride 0.2%
Leaf rust	Teak, sisham, ber	Spraying 0.2 % Zineb Spraying 0.2% wetable sulfer
Leaf blight	Neem, Eucalyptus,	Spraying Carbendazim 0.2%
Powdery mildew	Teak, neem, Casuarina	Spraying 0.2% Dithane Z-78 or Bordeaux mixture 0.1%

Biological Control Measure

Biological control is defined as the use of a living organism to control or manage another living organism. Natural enemies include parasites, predators, fungi, nematodes and viruses. Most biological control programs are directed at insects and mites. However, plant disease suppressive composts are being used to effectively reduce problems with certain root rot pathogens. In addition, new products are being introduced for biological control of root and foliar diseases.

Natural enemies are often used similar to a regular fungicide application program; that is, introduced on a regular basis. Even more than with fungicide, disease management will fail if pest populations are too high when the natural enemies are introduced. Biological controls are not rescue treatments. Often, disease pathogen populations will not decline and probably will even increase for several weeks to two months after introducing natural enemies.

Integrated Nursery Disease Management

- Selection of apparently healthy seeds/propagules for seedling production.
- Seed dressing with 0.2% Carbendazim/Methyl thiophanate/ Benomyl/Thiram.

- Sowing in sterilized/fumigated, clean beds and adequate watering.
- Using sterilized budding knife, secateurs, and scissors during budding and grafting.
- Transplanting seedling after root dip for 3-5 min in 0.02% Carbendazim solution.
- Healthy planting material maintenance by keeping them under proper sunlight, watering and clean environment.
- Frequent examination of seedling health and removal of diseased stocks.
- Foliar spray of 0.2% Carbendazim/Dithane M-45 at regular interval.

Table 3. Important nursery diseases, symptoms/damage and their management

Disease	Symptoms and Damage	Management
Anthracnose	Regular to irregular spots onleaves with dark margins and grayish centre.	Spary Carbendazim/ Bitertanol/ Thiophenate methyl Kavach/Rovral (2 g/*l*.) in humid weather
Bacterialcanker	Cankerous spots on leaves	Apply Copper oxycholride @ 3g/litre + Streptomycin @ 5g/10 litre duringrainy season
Powderymildew	White mealy growth on leaves	Wettable sulphur @ 2g/litre Carbendazim/ @ 1.2g/litre Bitertanol/Calixin, Kavach/Rovral (2 g/*l*.) in September-March
Downymildew	Pale-olive spots with milkywhite mycelium on the undersurface of leaves	Apply Metalaxyl + Mancozeb should be apply at 3-7days interval and

		systemicfungicide at 10-15 days intervals.
Rust	Pale yellow pustules ofuredospore are prominent onunder surface showed necroticpin head lesions.	Spraying Chlothalonil (2 g/l) or Hexaconazole @ 1g/l) are recommended
Damping off	The infections take place at thebase of the young stems or at thesoil level. Tissue becomes water soaked and rapidly collapse thustopping the seedlings. These pathogens cause pre - and postemergence damping off and wirestem of seedlings. It causes mortality of seedlings.	Treat seed/soil/ media with Captan/ Soil application *T. viridae* @ 2.5 kg + 10 kg FYM (keep for 7 days) & apply in yield. Seed treatment with Ridomyl Gold @ 2.5g/kg seed (Apron) thirum/ Tricoderma etc.
Stem rot/footrot/ collar rot	Rooting of seedling stem nearcollar region	Spray Kavach (Chlorotholonil @ 2g/litre) Rovial/ Metalaxyl @ 1g/litre Mancozeb (Iprodione 2g/litre)
Leaf spot	Dark brown to light brown, tinypustules on leaf surface, thepustules soon enlarge and turnb -lakish in the centre	Spray Carbendazim/ Bitertanol/ Thiophanole methyl Calixin,Kavach/ Rovraal (2 g/*l*.) in rainyseason/ November
Wilt .	The foliar are yellowing andproduction of crook -neck shoots.The leaves and shoots wither and become brownish. Stems whencut open show browndiscoloration at the vascularregion.	Soil fumigation and treating thenursery beds with benlate or withThiophanate soil application methyl and using *Tricoderma* etc.

Weed Management in Nursery

Weed management is an important component of plant protection for improving the production potential of nursery. It includes management of the weeds in such a way that the nursery sustains its production potential without being harmed by the weeds. Weeds are the plants that grow without human efforts and are not wanted. They grow in the fields where they compete with nursery plant for water, soil nutrients, light, and space. Proper weed management is a pre-requisite for obtaining higher input efficiency. Weeds also act as alternate hosts that harbor insects, pests and diseases and other microorganisms. Some weeds release growth inhibitors or poisonous substances into the soil. These may be harmful to the nursery plants, human beings and livestock. Weeds reduce the quality of marketable nursery produce, leading to increased expenditure on labour, equipment and chemicals for their removal.

Weed management is done through the mechanical, cultural and chemical means. Use of biological control methods in nursery is being considered, but still not much in use. Use of herbicides is an important method of weed-management technology. New hand-tools and implements have also been designed to assist in weed management in nursery.

Nursery plants are exposed to severe competition from weeds. Most of these weeds are self-sown and they provide tough competition to the nursery plants due to their faster growth than the growth of nursery plant in their initial stages.

Control Measure

Control practices reduce or suppress weeds but do not necessarily result in the elimination of any particular weed species. Weed control, therefore, is a matter of degree and depends upon the goals of the people involved, effectiveness of the weed control tactic used and the abundance and tenacity of the weed species present. There are 4 general methods of weed control: Physical, cultural, biological and chemical.

A. **Physical Methods:** Hand pulling or hand weeding, hoeing, tillage, mowing, burning, flooding, smothering etc. are examples of physical methods of weed management. These methods involve the use of physical energy through implements either manually or bullock drawn or power operated. Nurseryman mostly resort to hand weeding with the help of hand chisel (khurpi), hand hoe, spades, etc. It is one of the most commonly used methods but is back breaking, time consuming and costly effort.

B. **Cultural Methods:** Weeds are better competitors than nursery plants for light, water, nutrients and space. However, good nursery practices manifest the conditions so as to enable the nursery plants to compete with the weeds successfully. Such practices can reduce the interference of the weeds to the minimum and optimize the nursery production. Quality seed with good germination will give the nursery seedling a vigorous and close stand, and would enable the seedling to steal a march on the weeds. Varieties well adapted to a season will complete better with the weeds than those poorly adapted to it. Selection of crops which are quick growing and short duration varieties with larger leaf area and good branching or tillering ability is essential to compete with the weeds. Proper seed rate, depth, time and method of sowing, and the use of the most appropriate method of irrigation and manuring can also minimize the losses due to weed infestation.

C. **Biological Method:** In this method, the natural enemy of a weed plant is used to control the weed. The requirements for the success in this method are:

1. The weed species must have been introduced and in the process of introduction must have been freed from its natural parasites or predators.

2. The natural predators and parasites must be introduced to prey upon or parasitize the weeds but they in turn must be freed from parasites in order to carry out their work for destruction unhampered.

3. The destructive agents must be highly specialized so they these are able to thrive even under starvation condition on agricultural plants of the new habitat. Root-borers, stem-borers and internal seed-or fruit feeders are more highly specialized than the foliage feeders.

D. **Chemical Method:** Weed management in horticultural nursery constitutes a major input in their production of seedling. Losses due to weeds are highest in least developed crop production systems and lowest in most highly developed ones. In tropical and subtropical countries, 70% of the labour input is diverted for weeding. A broad spectrum of monocot and dicot weeds infests these crops.

Integrated Weed Management

Integrated Weed Management (IWM) approach aims at minimizing the problem of residue of herbisides found in plant, soil, air and water. An Integrated Weed Management may be defined as the combination of two or more weed control methods at low input levels to reduce weed competition in a given cropping system below the economical threshold level. It has proved to be a valuable concept in a few cases, though much is still to be done to extend it to the grass root farmer level. An IWM involves the utilization of a combination of mechanical, chemical and cultural practices of weed management in a planned sequence, so designed as not to affect the ecosystem.

The nature and intensity of the species to be controlled, the sequence of nursery plants that are raised in the rotation, the standard of nursery plants to be produced, the time of availability of the resources used in the method and the economics of different weed-management techniques are some of the potent considerations that determine the success for the exploitation of the IWM approach.

Control Measures of Weeds Infesting Fruit, Vegetable and Flower Nursery Seedling.

Crop	Recommended herbicide
Mango	Diuron or Oxyfluorfen
Banana	Diuron/Simazine/Oxyfluorfen Paraquat/ Glyphosate
Pineapple	Bromacil + Diuron
Grape	Oxyfluorfen or Diuron
Papaya	Fluchloralin or Butachlor or Alachlor
Citrus	Diuron
Guava	2, 4-D or Paraquat or Diuron
Strawberry	Simazine or Chloroxuron
Cocoa	Basta

Coconut	2,4-D or Ester or Diuron or Glyphosate
Coffee	Oxyfluorfen or Glyphosate
Oil palm	Gramuron (Paraquat+Diuron)
Cabbage	Oxyfluorfen or Fluchloralin
Cauliflower	Oxyfluorfen or Pendimethalin
Tomato	Alachlor or Fluchloralin
Brinjal	Fluchloralin
Chilli	Alachlor or Butachlor or Fluchloralin
Onion	Fluchloralin
Cucurbitaceous crop	Alachlor or Butachlor or Fluchloralin
Rose	Diuron or Oxyfluorfen orAtrazine
Gladiolus	Fluchloralin
Chrysanthemum	Oxyfluorfen
Marigold	Simazine
Carnation	Oxadiazon

Biopesticides Application In Nursery

Biopesticides are certain natural plant products that belong to the so called secondary metabolites that include thousands of alkaloids, terpenoids, phenolics and minor secondary chemicals.

Biopesticides have usually no known function in photosynthesis, growth or other basic aspects of plant physiology; however, their biological activity against insect pests, nematodes, fungi and other organisms are well documented. Every plant species has developed an inbuilt unique chemical complex structure for protection against pests. The plant kingdom offers us a diverse array of complex chemical structures and almost every imaginable biological activity. These biodegradable, economical and renewable alternatives are used especially under organic farming systems.

Crop losses due to insect pests and diseases are estimated to be more than 30 per cent in terms of agricultural crop produce. Though the crop productivity increases significantly due to use of mostly chemical pesticides, their continuous use contributed to increasing environmental threat to natural resources, wildlife, non-target beneficial species. It also results in development of pest resistance and resurgence of the pest attack. Mounting concern for environment has, therefore, led to evolve an alternate coherent pest management programme which is eco-friendly, self sustaining and yet provide effective management of pests and diseases. In this context the use of biopesticides or pesticides of microbial origin are becoming increasingly important.

The Characteristics of Biopesticides

- Eco-friendly,
- Possess higher degree of host specificity,
- Genetically stable and do not pose the risk of resistance development,
- Compatible with biofertilizers and many agrochemicals,
- Can be exploited under the ambit of integrates pest management (IPM).

Viral Biopesticides

Viral biopesticides of baculovirus group namely Nuclear Polyhedrosis Viruses (NPVs) and Granulosis Viruses (GVs) offer great scope as crop protection agents on high value crops such as cotton and vegetables against lepidopteran pests like *Helicoverpa armigera* Hbn. and *Spodoptera litura* Fabr. More than 500 *baculovisurses* (BVs) have been reported so far. Some of the insect pests from which occlusion viruses have been reported in India have great potential in IPM.

Bacterial Biopesticides

Many spore forming and non-spore forming bacteria are known to be effective against a wide spectrum of insects and pests. The crystalliferous *Bacillus thuringiensis* (Bt) has been found to be effective against several species of lepidopteran pests. Its insecticidal activity is primarily caused by parasporal crystal (delta endotoxin) produced during sporulation. The discovery of isolate capable of acting against coleopterans (*B. thuringiensis tenebrionis* and *B. sphaericus* against mosquito and *B. moritai* against dipteran insects has further extended the scope of bacterial insecticides. Several bacteria like, *Pseudomonas fluorescens, Bacillus subtilis, Streptomyces nigrifaciens, Agrobacterium radiobacter* and *Azotobacter spp.* are known to be potential bio-agents against several plant pathogen. Most of these bio-control bacteria are, however, in the stage of empirical application, either in the green house or in the field with a few exceptions. There are reports of bacterial bio-agents in enhancing the growth and greater root volumes of crop plants when applied as seed treatment or soil application. This envisages that the responses may be as biofertilizer or by the control of some undiagnosed plant pathogens.

Fungal Biopesticides

Fungi unlike bacteria or virus do not require ingestion for infection; so sucking pests are also targeted by primary contact or by secondary uptake from sprayed vegetation. The pathogenesis begins with germination of conidia on the cuticle and penetration and development inside the host leading to death of host essentially under high humid condition.

Popular Biopesticides used in Nursery for Pest and Disease Management

Microorganism	Target pest/disease
Insecticides	
Bacillus thuringiensis (Dipel)	Caterpillar, larvae
B. thuringiensis isrealensis (Acrobe)	Mosquito, Black files
B. thuringienis tenebrionis (Novodor)	Colorado potato beetle
B. thuringiensis aizawai (Certain)	Wax moth, Diamond black moth
B. Spharicus (Spic biomass)	Dipteran insects
B. papillae (Doom)	Coleopteran insects
B. moritai (Labillus)	Dipeteran insects
Verticillium lacanii (Vertalee/Biolin)	Aphids, scale insects, mite, thrips
Entomopthors	Aphids, mites
Numurea releyi	Lepidopteran larvae
Hirsutella thompsonii (Mycar)	Citrus rust mite
Beauveria bassiana (Bovarin/Blorin)	Colorado potato beetle, Codling moth, Coffee, berry borer
Asohersonia aleyrodis	Citrus scale insect
Metarhizium anisopilae (Metaquine)	Spittle bug
Paeciliomyces lilacinus (Bicon)	Nematodes
Metarhizium flavoviride	Locust, grasshopper

Fungicides	
Trichoderma sp. (BINAB-T)	*Chondrostereaum purpureum*
T. harzianum (F-stop)	*Pythium sp.*
Trichoderma sp. (Trichodermin)	*Botrytis, Pythium, Sclerotinia, Verticillium sp.*
T. harzianum 9T-39) (Trichodox)	*Botrytis cinerea*
T. harzianum (Tricho dowels)	*Rhizoctonia solani,Pythium sp.*
T. viride (Antagon TV)	*R. Solani, Macrophomina phaseolina*
Tt. viride (Sun-Derma)	*R. Solani, Macrophomina phaseolina*
Trichoderma sp. (Ecofit)	*R. Solani, Macrophomina phaseolina*
T. virens (Gliogard)	*Pythium ultimum*
T. virens (Soilgard 12 G)	*Pythium sp.*
Amphelomyces quisqulais	*Saphaerotjeca fuliginea*
Pichia quiliermondii	*Penicillium sp.*
Aspergillus niger AN-27 (Kalisena)	*Pythium, Fusarium, Macrophomina,Rhizoctonia, Sclerotinia sp.*
Pseudomonas flurescens	*Rhizoctonia solani, Scleritium rolfsii, Fusarium oxysporum, Legaeumannoyces graminis var. tritici*
Agrobacterium radiobacter	*Agrobacterium tumefacience*
Bacillus subtilis	*Fusarium udum, Fruit Rot of avocado,Xanthomonas malvacearum, Rhizoctonia solani*
Nematicides	
Nematode trapping fungi Bacteria	*Myrothecium verrucaria, Paecilomyceslilacinus, Bacillus firmus, Pasteuria penetrans*
Mollusc parasitic nematode	*Phasmarhabditis hermaphrodita*
Weed control	
Fungi/ Bacteria	*Colletotrichum gloeosporioides, Chondrostereum purpureum, Cylindrobasidiumleave Xanthomona campestris pv. poannua*

Advantages of Biopesticides

- Biological control methods can be used as part of an overall integrated pest management (IPM) program to reduce the legal, environmental, and public safety hazards of chemicals.
- It may be a more economical alternative to some insecticides.
- Some biological controls are often very specific for a particular pest. Other helpful insects, animals, or people can go completely unaffected by their use.
- There is less danger of impact on the environment and water quality.

Disadvantages Biopesticide

- It takes more intensive management and planning.
- It takes more time to show measurable effect;
- It requires more record keeping, more patience, and sometimes more education or training.
- Successful use of biological control requires a greater understanding of the biology of both the pest and its enemies.
- In some cases, biological control may be more costly than pesticides.
- Often, the results of using biological control are not as dramatic or quick as the results of pesticides use.

❑❑❑

9

Plant Propagation Structures

The propagation of crops is a natural process that is undertaken under modified and controlled environments in order to enhance this natural process. These environments are modified to ensure the optimum growth of the crop in question. Propagation structures are established to create environments that can be modified and controlled. Facilities required for propagating plants by seed, cuttings, and grafting, and other methods include two basic units. One is a structure with temperature control and ample light, such as a greenhouse, modified quonset house, or hotbed—where seeds can be germinated, or cuttings rooted, or tissue culture microplants rooted and acclimatized. The second unit is a structure into which the young, tender plants (liners) can be moved for hardening, which is preparatory to transplanting outdoors. Cold frames, low polyethylene tunnels or sun tunnels covered by Saran, and lathhouses are useful for this purpose. Any of these structures may, at certain times of the year and for certain species, serve as a propagation and acclimation structure.

In general, propagation is done in various structures including:

- Germination rooms
- Greenhouses
- Shade-houses
- Specialised tissue culture laboratories

Germination Rooms

During the propagation of tree crops, the rootstocks are usually propagated from seed. Seed trays that have been sown with rootstock seed are kept in germination rooms where they remain until the seedlings have developed and are ready to be transplanted into seedling trays. Germination rooms are usually permanent buildings, with constructed walls that are plastered and painted. The standard size for a germination room is 3m x 2m with a ceiling height of 2.5m. A germination room is developed in a way to retain as much heat as possible with limited air-circulation and the potential for build-up of humidity. The rooms are generally equipped with fluorescent lights as a source of red light, which promotes germination.

Greenhouses

Most greenhouse structures are manufactured from either shade cloth (shade houses) or plastics sheeting.

Shade houses

Shade houses are green houses made from shade cloth. It is the cheapest system available and is easy to construct. Shade cloth provides protection against sun, wind, hail and rain. The shade cloth does not absorb water and does not decay as easily as plastics sheeting. Shade cloth allows wind to circulate into green houses, assisting in cooling of the construction. It will provide protection against animals, large insects and birds but will not protect it against small insects. Temperature and humidity cannot be regulated either.

Plants grown under shade cloth are reasonably protected from harsh environmental conditions. Aeration is passive as the wind speed is reduced. The main concern is the transpiration rate. With extreme light-intensity, the temperature in the shade-house increases and the relative humidity decreases. Keeping cells turgid becomes the main goal. The choice of shading percentage is influenced by this goal. The ultimate is an environment where light-intensity is reduced, which in turn reduces the temperature and increases humidity.

All shade houses are equipped with irrigation and fertilisation systems with fixed access points for pest and diseases control systems.

Shade netting

Shade netting is produced from knitted polyethylene cloth. The product is available in green, black and white and has density ranges from 20 to 90 %. The density indicates the percentage Ultra violet light that is absorbed by the netting. Shade cloth is available in different colours and shading percentages. Depending on the area, the percentage of shading varies between 30% and 60%. In the flower industry a pink net with 40% density netting is used to stimulate flowering.

Types of shade houses

Shade houses are constructed either as flat roofed or pitched roofed shade houses.

1. Flat roof shade houses

Flat roof shade houses are used in areas where low or no hail is expected. They are easy to construct and cheaper than pitched roof constructions. The construction is laid out so that plant rows are north - south orientated. The areas for one flat roof shade house should not exceed 100 X 100 m. The area is measure off and divided into sub-blocks each 3 X 6m or 6 X 6m. This is the spacing at which the carrier poles will be planted. Poles are planted 0.5 to 0.9 m deep in the ground. The height of poles dependent on the structure required. The poles used are usually treated gum poles. The perimeter poles are anchored by using steel cable or wire. The anchor-wire is anchored to a stay or a specialised anchor which is driven into the ground before the anchor wire is secured. Stay wires are pulled across the top of the planted poles and the shade netting secured on top of the wire grid that is created. The netting is secured to perimeter poles using high tensile steel wire or cable pre-fitted through the eyelets in the netting. The netting is now fasted to the cross wires at 1 m intervals.

Pitched roof shade houses

Pitched roof structures are suitable for areas where hail is prominent. The construction uses more materials and longer poles than a flat structure and is therefore more costly to build. The standard height for construction poles are 4.2 m and 2.6 m. The shorter poles are planted at 2m intervals on the outsides of the structure. The longer poles are planted on the other side and netting secured over the construction.

Plastic clad greenhouses or tunnels

Plastic clad greenhouses and tunnels consist of pole structures covered with specialised plastics cladding. The advantages of plastics tunnels include:

- The tunnel can be sealed, aiding in proper climate control.
- Crops are not damaged by environmental conditions.
- Use is not limited by season.

The disadvantages include:

- Plastics sheeting has a limited life-span
- Construction cost are high
- Temperature control units must be installed especially to control extreme variations in temperatures.

Tunnels are constructed as a series of sequential metal arches, which are covered with plastic sheeting. The system is usually fitted with a door and cooling system. The sizes of the tunnels differ, depending on the manufacturer or supplier.

Selection of Green house type

The materials used to build a greenhouse depend on the environment in which the structure will be built as well as the environmental requirements of the plants. If the greenhouse is to be built in a location that is known to be windy, provision must be made to strengthen the structure to withstand wind. If the structure is to be built in conditions that are generally humid, special corrosion-resistant material must be used. At the same time, the design and material used in the greenhouse must contribute in creating the ideal environmental conditions that promote plant growth.

Greenhouses generally have a metal framework (aluminium or galvanised steel or stainless steel) that is rust and corrosion resistant. The framework is clad with ultraviolet (UV) resistant polyethylene sheeting. Initially glass was (and still is) used to cover the metal framework, but it is very expensive. A variety of different plastics cladding can be used, but the best results in terms of lifespan and light penetration, have been obtained with 100 micron UV resistant cross-woven polyethylene sheets.

Greenhouses provide a more favourable environment than the outdoors. The propagator is not at the mercy of nature's extremes in as far as temperature is concerned and propagation and plant production can take place year-round. If environmental control systems are installed, temperature is controlled to allow normal respiration, even when photosynthesis is reduced as is commonly found when cloudy weather prevails.

Humidity ensures that the environment is cooled. This reduces the respiration rate resulting in fewer carbohydrates being used (burnt) for respiration and thus more being available for plant growth. Greenhouses are equipped to humidify the air when required.

The following environmental control equipment may be installed:

- A wet-wall is constructed by using a honey combed cellulose pad, a distribution pad, perforated PVC pipes that provide the water that run through the wall, a pump and a shallow tank - used for cooling;
- A set of extractor fans, or exhaust fans, that draw cooled and humidified air through the greenhouse;
- An irrigation system for irrigating the plants;
- An automated control system that regulates the light intensity or day length through curtains or supplementary light.

Air movement also helps to avoid injurious levels of temperature and CO_2 . The propagator must constantly check the balance between heat build-up and the availability of CO_2 for plant growth.

Greenhouses to Optimize Seedling Production

The principal role and function of greenhouse facilities is to modify or manage environmental conditions to optimize plant health and development. Although greenhouse structures serve many purposes, from producing transplants, to in-ground production of high value crops, to early and late season extension in a range of climates, this lecture focuses on using greenhouse facilities for seedling production.

1. Optimizing ***germination***: Propagation structures, combined with the knowledge and experience of the greenhouse grower, can be managed to create optimal environmental conditions (e.g., temperature, air circulation, light, and soil medium moisture) that facilitate rapid germination and early crop establishment.
2. Promoting healthy ***early seedling development***: Ongoing management of environmental conditions (temperature, air circulation, and moisture delivery) is required as seedlings develop, but with most species, seedlings' physiological tolerance

expands and precise environmental control may be less necessary to maintain optimal development. When greenhouse space is at a premium, young seedlings are typically moved to alternative structures (from a greenhouse to a hoop house, for example) to make way for the next generation of crops most dependent on precise environmental control.

3. Managing ***seedling maturation and hardening off***: Mature seedlings will typically have a balance of root and shoot growth—at least two sets of true leaves and an ample root system that holds together the root ball when removed from the growing container. "Hardening off" is the final step in preparing seedlings for transplant and uninterrupted growth. In the final 3–10 days in the greenhouse zone, seedlings should be outdoors and exposed to conditions that most closely resemble their future home in the ground. This acclimatization process reduces transplant shock, which can occur when seedlings experience an abrupt transition from the protected environment of the greenhouse to the less predictable conditions of a garden or field setting.

Passive and Active Environmental Management

The methods and tools used to manage environmental conditions in greenhouses normally fall into one of two broad categories: Passive or active. Here we briefly describe these categories;

1. Passive methods

Passive methods of environmental control are part of the functional design of most greenhouses and represent a low-tech approach that does not involve the ongoing use of energy to regulate conditions. They include:

a) Heating by the capture of solar radiation as sunlight passes through the greenhouse glazing and warms the interior air.

b) Cooling via side and end wall vents that draw in cooler air from outside and ridge vents at the top of the greenhouse that allow the heated air to be exhausted out of the structure. Shade cloth or whitewash can also be used to help cool the greenhouse.

c) Air circulation via the venting system. As with cooling, exterior air enters the structure through open side and end wall vents, and the air already in the greenhouse exits primarily through ridge vents and vents placed high on end walls.

d) Irrigation can be delivered by hand or by overhead spray systems.

e) Lighting comes exclusively from sunlight. Light reduction via whitewashing and the use of shade cloth is another form of passive management.

f) Additional physical methods to heat and cool the greenhouse include the use of shade cloth, white washing, and energy curtains.

2. Active Methods

Active environmental controls use an external energy source to power heating, cooling, venting, supplemental lighting, irrigation, and climate control systems. Active control mechanisms are not a substitute for passive methods, but rather are complimentary tools that allow growers to more precisely and predictably create desired conditions. Active methods include:

a) Heating via conduction (direct contact with heating source), convection (via warm air circulation), and radiant heat sources.

b) Cooling via evaporative mechanisms (pad and fan systems), swamp coolers, and fog systems.

c) Air circulation via exhaust and horizontal airflow fans.

d) Supplemental lighting, including incandescent, fluorescent, high intensity discharge, and high pressure sodium lights.

e) Automated irrigation systems, including overhead sprinklers.

f) Thermostats, stage controllers, and computer-directed environmental controls that monitor and control various heating, cooling, circulation, lighting, and irrigation systems.

Environmental Control in Different Types of Propagation Structures

1. **Passive Solar Greenhouse**: Good environmental control is possible in relatively lowtech facilities, especially in milder climates where growers do not face extremely hot or cold conditions. As described above, these greenhouses rely on passive techniques.

   a) Trap solar radiation to warm the air and thus the crops.

b) Cooled through the use of venting systems: Combination of end wall vents, roll up/down sides, and ridge vents, to draw in cooler external air, and exhaust warmer internal air.

c) Air circulated by manual control of inlet and outlet venting; complements the heating and cooling capacity of passive solar design.

d) Moisture regulation/delivery managed by the grower through manual/hand delivery, semi-automated, or automated delivery systems.

e) Microclimatic heating may be possible; offers additional environmental control when power is available to supply the system, but this goes beyond the purely passive.

2. **Enclosed (semi) Automated Greenhouses**: Precise environmental control achievable via active mechanisms:

a) Characterized by ability to fully close growing environment, regulate temperatures/air circulation through passive and active venting and fan-driven air movement, heating/cooling by fans, furnaces, swamp coolers, etc.

b) The interplay of environmental conditions and sophistication of active management tools dictate the precision of environmental control: More precise control comes with more responsive systems and lesser extremes in conditions to be regulated.

c) Trapped solar radiation works in concert with active heating mechanisms to create desired warmth.

d) Cooling via passive venting systems works in concert with active cooling mechanisms to create desired temperatures.

e) Microclimatic heating in root zone via hot water pipes or electric cables is often used to optimize conditions and speed the rate of plant growth.

f) Automated or manually controlled sprinkler systems are the norm in more actively managed and infrastructure-intensive greenhouses, but spot watering by hand remains critical to optimize plant health.

3. **Open Hoop Houses/Quonset Huts**: Can partially modify environmental conditions and improve plant health, especially in milder climates.

   a) Temperature modification: Umbrella-like coverage creates slightly warmer day and night conditions, which favors more rapid development than possible outdoors.

   b) Provide some buffering against effects of wind, though air circulation may be limited by how structure is located relative to prevailing winds, nearby windbreaks, and other structures, unless the hoop house is outfitted with roll up or roll down sides and end wall venting.

   c) Grower controls moisture regulation, delivering necessary irrigation through same means as in greenhouses.

   d) Can be used as an intermediate step between greenhouse and outdoors: Greater exposure of plants to wind and day-night temperature fluctuations offers a gradual step in the hardening off process, especially when the favorable conditions created in the greenhouse are dramatically different than prevailing outdoor conditions.

4. **Germination Chambers**: Small-scale, self-contained facilities that provide optimal control over temperature and humidity to facilitate rapid, high-percentage germination

   a) Whether pre-assembled or home made, chambers are comprised of a water-holding pan, a submersible heat source with variable temperature control, a water supply, insulated walls, suitable shelving to hold propagation trays, and doors large enough to easily move containers in and out of the chamber. Units usually hold between 24–48 propagation trays but can be designed larger or smaller to meet needs.

   b) Together, these elements create an energy efficient, very consistent environment for germination where humidity and temperature can be optimized to greatly improve and speed germination.

   c) Rapid germination in chamber's compact space can save growers significant heating or cooling costs, depending on the season.

d) Higher percent germination can reduce seed costs.

e) Must be regularly monitored: Once seedlings germinate, propagation trays must immediately be removed from the darkness of the chamber and moved to a sunny location to facilitate normal cell development and photosynthesis.

5. **Cold Frames**: Small-scale, low-tech structures used to modify environmental conditions. Similar to passive solar greenhouses and hoop houses that do not contain any supplemental infrastructure.

   a) Simple structures, normally placed directly on the ground, consisting of four low walls (insulated in colder climates) and a sloping, hinged roof that allows in sunlight for warmth and photosynthesis. Roof can be opened to facilitate cooling and air circulation.

   b) Often built of recycled materials, such as wood and rigid insulation board for siding and old windowpanes or clear acrylic panels salvaged from local resource recovery facilities for the roof.

   c) Sunlight can quickly heat internal air and enclosed air mass will provide some buffer against cold nighttime temperatures; due to small size, cold frames provide limited buffering capacity and are prone to rapid temperature shifts as external temperatures change.

   d) Cooling and air circulation are achieved through opening the roof. Relying on this passive exchange, while effective, can be problematic if the cold frame is left closed for too long during warm conditions.

   e) Irrigation is usually done by hand and can provide a secondary form of cooling through evaporation.

   f) Through greater venting or leaving a cold frame open overnight, increased air flow and day-night temperature fluctuation can help initiate hardening off.

6. **Outdoor Benches**: In most growing environments, seedling maturation or hardening off is completed by placing seedlings outdoors on benches, exposing plants to conditions that closely approximate the in-ground environment they are moving

toward. As detailed above, full exposure to sunlight, wind, and temperature swings stimulates carbohydrate reserve buildup and strengthens cell walls so that plants can withstand the vagaries of the in-field environment.

7. **Shade Structures**: While most annual vegetable seedlings require full sun to optimize growth, shade cloth may be needed in hotter climates to prevent soil media from drying out so rapidly that the grower must constantly devote time and water to keep young seedlings healthy.

Irrigation Management and Delivery

Greenhouse irrigation concepts and terminology are similar to those used in the garden and field setting. However, because of the small soil volumes plants are growing in and because of the design of propagation and nursery containers, water behavior in greenhouse containers and consequent practices can be quite different.

a) **Saturation:** As with field soils, saturation in containers comes at the point when irrigation water fills all of the pore space in the soil medium, but in high quality mixes, this is only a very temporary state and excess moisture quickly drains from the mix.

b) **Container Capacity:** Similar to Field Capacity, container capacity is when excess water has drained, air has returned to part of the pore space, and maximum water is held in the pore spaces against the forces of gravity.

c) Percent Container Capacity describes the relative availability of water in the mix as water is lost to uptake by the plant roots and evaporation.

d) Perched Water Table describes the water that is held at the bottom of flat-bottomed containers. This condition may be detrimental to plant health if the soil mix does not contain an adequate amount of coarse materials to promote good aeration throughout the root zone.

e) Percent Surface Dry Down is another important concept in the greenhouse setting, and applies specifically to the germination phase of seedling production when frequent but small quantities of water must be delivered to facilitate germination and prevent desiccation of newly emerging roots.

- For the vast majority of seed-grown crops, a small quantity of water should be reapplied when somewhere between 30–50% of the visible soil surface has dried down.
- For larger-seeded crops such as sunflowers and members of the cucurbit family, growers typically allow 100% of the surface soil to dry down before re-applying moisture.

f) **Post germination:** Greenhouse growers deliver water in direct response to crops needs, the age and stage of development of their crops, and the immediate and anticipated environmental conditions that crops are experiencing.

Automated and Semi-Automated Climate Control Infrastructure

1. Thermostats are the least expensive and unfortunately least accurate environmental management tools available for automation.
   - Ther mostats are positioned in the greenhouse to turn heating and cooling equipment on at pre-determined temperature thresholds, based on your climate and crop needs.
   - Older thermostats, while convenient, are notoriously inaccurate and may not give the precise control desired. Additionally, separate thermostats are required to control heating and cooling equipment.
   - Modern thermostats, while more expensive, offer more accurate control by using digital or electronic technology to monitor temperatures and operate environmental control equipment.
2. Stage controllers offer dramatic improvement in precision of environmental control by linking the operation of heating and cooling devices, air circulation mechanisms, and even the deployment of energy curtains and shade cloth in the higher-end models.
   - Stage controllers typically provide one or two set points to activate the heating mechanisms in your greenhouse and three, four, or more stages of activation of the cooling mechanisms. For example, as the greenhouse heats beyond a given threshold, at stage one vents will open. With continued heating, vents and fans will operate and with further heating, the pad and

fan cooling system will be activated to maintain the desired temperature.

- Stage controllers are only limited by the range of set points that activate your heating and cooling mechanisms; higher-end models may also offer a data recording feature.

3. Computer directed environmental controls offer the maximum level of precision in the total environmental control of the greenhouse.

    - Computer zone controllers, with or with without a PC, links all aspects of environmental control-temperature, air circulation, water delivery, and lighting through a single device, offering a high degree of flexibility and the ability to customize the system.
    - Integrated computer systems, using a PC, also provide maximum control of all aspects of environmental conditions, with the added advantage of being able to control multiple zones or separate structures from a single device. The most sophisticated systems offer remote access through smart phones, tablets, and laptops, as well as the ability to receive alarm warnings whenever conditions are out of the desired range or components are malfunctioning.

Potential Problems with Structures

Structures are used to provide an environment that is and remains easily controllable. Should any structure lose this property, it becomes less useful to the propagator and economic losses may occur.

Potential faults may occur with the control devices, such as probe malfunction, the wrong interpretation of information that has been relayed by the instrument or the structural frame itself.

Propagation structures are generally held together with links and joints which can loosen. Therefore, links, joints and the generally stability of the structure must be checked regularly.

Cracks or cuts in the covering material, whether it is the polyethylene sheeting or shade-cloth, results in unwanted changes of the environmental conditions within the structure. Plants are exposed to harsher conditions with an increased probability of damage to the plants.

Greenhouses are at greater risk. Fans and wet-wall malfunctions have disastrous implications. Temperatures can quickly reach levels where plants cannot continue normal metabolic processes, resulting in severely wilted plants which are the result of low photosynthesis rate. Poor aeration can lead to a drop in the CO_2 level, limiting photosynthesis. High respiration rates are required with a constant supply of O_2 and CO_2.

The selection of the wrong covering material for a shade-house can result in poor light intensity during a specific time of the year. Light is filtered before reaching the plants. The location and season are important factors that determine the quantity of light received. Light intensity is generally high from October to March, peaking in January.

❑❑❑

10

Growing Media

Since the growth medium relates to every cultural practice in the production of nursery crops in containers, the selection and preparation of the medium is extremely important and will pay great dividends in terms of plant growth and quality. Several combinations of media with desirable physical, chemical and biological properties can be used but the goal should be consistency from batch to batch. The best soil to use will depend on the species and what is available. In general a sandy loam texture and moderate to slightly acidic reaction (pH 5.5 to 7.0) represents the most favourable condition for forestry, horticulture and agroforestry nursery stock although slight variations may be necessary for certain species. In very acidic or alkaline soils certain plant nutrients are either leached or become insoluble and therefore unavailable. Acid soils below a pH of 4.5 are low in exchangeable calcium and magnesium and lack most of the important plant nutrients like nitrogen, phosphorous, potassium, sulphur and some trace elements (copper, zinc and boron) with only iron and manganese being available.

For the above reasons it is always advisable to know the basic chemical characteristics of the soil being used for germination and transplanting. To avoid having to take a test with every production cycle, it is important to be able to obtain soil from the same source or soil type on a steady basis but this is not always possible. Although the sifted sand used in germination beds and container media appears clean and sterile it should be washed carefully to remove dust, harmful dissolved substances and weed seeds before use.

Raising and lowering the pH value: The pH value of soils can be raised by adding lime or a fertiliser that reduces acidity such as sodium nitrate or calcium nitrate. An alkaline or neutral soil can be changed by the application of acid forming fertilisers such as ammonium nitrate and urea. Care must be taken with these fertilisers not to over fertilise the seedlings.

Growing Media for Propagation and Germination Beds

The physical structure of the medium in which seeds are germinated is crucial both for germination and early seedling establishment. The following growing media characteristics favour the growth of tender young seedlings:

1. Good aeration which permits an adequate supply of oxygen to the root system.
2. Good texture to facilitate contact between the seed and the growing medium. Large spaces between the medium particles should be avoided.
3. Little physical resistance so that the emergence of the seedling is not restricted and root penetration is fairly easy.
4. Infiltration capacity that permits easy watering and avoids crusting on the surface as often occurs with fine sand and silt.
5. Absence of fungi, bacteria, nematodes and weeds.

In view of the above, growers favour a medium loam texture. Incorporation of peat, loose sifted soil, soil/sand mixes or any of the imported specially prepared soil mixes may be necessary to achieve the desired structure.

Propagation media are a major component of a propagation operation as they host the roots that feed and stabilise the plant. Most media are invented to meet specific requirements and can therefore be considered (to a certain extent) as artificial, as opposed to natural.

In order for a plant to feed easily, the medium must retain and make available enough water and nutrients. Technically the ability of a medium to retain nutrients is referred to as the cation exchange capacity (CEC) of the medium. Media that are dense, with small pores, inhibit water movement and make it more difficult for the plant to take up nutrients. The other extreme, is an excessively porous medium which will be closer to natural, coarse sand. Water will leach most nutrients from such a porous medium, making nutrients unavailable to plants. Where only a porous medium is available, slow release fertiliser could be used in conjunction with closely monitored water application.

Seedlings have to be able to develop a root system that will later sustain the plant throughout its entire lifespan. The medium has to provide as little resistance as possible to the developing roots. However, the medium must be sufficiently firm to hold seedlings and cuttings in place. The range of media used includes:

- Compost
- Sawdust
- Vermiculite
- Perlite
- Peat moss
- Coconut fibre

Besides the considerations mentioned earlier, the propagator must take the following into account:

- Does the medium require any additional sterilisation? Can it be sterilised? At what cost?
- Is the salinity of the medium acceptable?
- Is the medium free of contaminants such as weeds, nematodes, pathogens and pesticides?

Fine sawdust is not recommended, because of its property to compact quickly and reduce pore-space.

Growing Media for Transplant Beds

For the production of bare root stock, the available nursery soil can be used if its quality measures up to the following points:

1. Good texture with limited clay content to avoid compacting and drainage problems.
2. Organic matter present to ensure optimum physical soil conditions.
3. Adequate nutritional level so that additional fertiliser application will not be necessary or minimal.
4. Absence of weeds and harmful pests and diseases.

In cases where the soil in the nursery is considered to be inappropriate, soils found in the forest or carefully selected top soil from some land use development projects should be made available. Care should be taken to avoid sub-soil which is often available in large quantities at construction sites. Before use, the soil should be screened through a coarse sieve with a 1 cm mesh. Figure 1 illustrates two simple methods of screening soil. After screening, the soil is mixed, if necessary with sand and/or organic matter and appropriate fertiliser if required, prior to bed formation. Rest the growing media for at least two weeks after preparation to allow unwanted weeds to germinate and be removed.

Growing Media For Container Seedlings

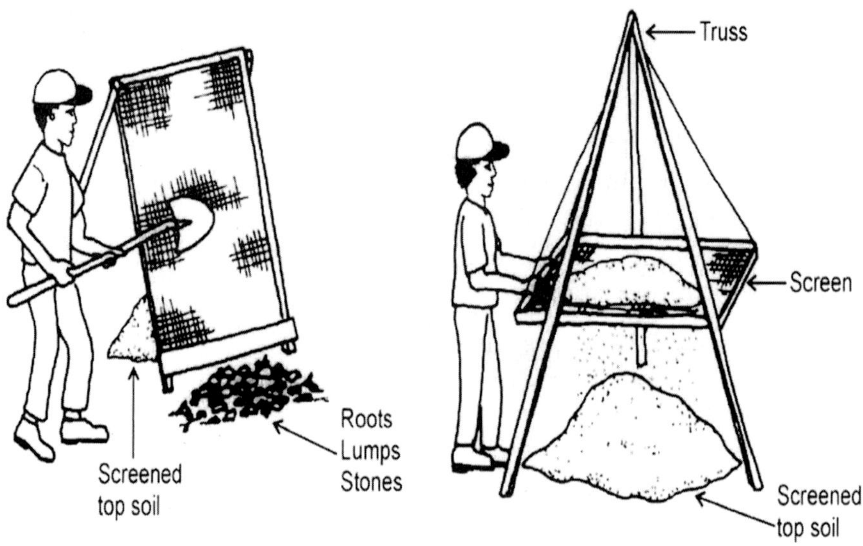

Fig. 1 Two Simple methods of screening soil

The soil for containers is similar to that for bare root beds as described above. It should be well drained and a mix of soil and sand with organic material is common practice to prevent the root ball from disintegrating when the container is removed at time of planting. Heavy clays are undesirable because of poor drainage and compaction on drying out, and if avoided, the practice of adding sand would no longer be necessary. Presently nurseries use a 3:2:1 mixture of top soil, organic matter and sand. When the supply of top soil is variable and organic matter often unavailable, the quality of seedlings produced leaves much to be desired. The goal should be a 50:50 mixture of top soil and organic matter, avoiding the use of sand.

Organic Material

The following range of organic material has been used irregularly in mixture with soil to produce the growing media for containers in forest and horticultural nurseries: animal manures, bagasse, chicken litter, coffee waste, coir dust, saw dust, sludge mostly in the form of compost or partially composted material, and commercial organic matter.

The use of one or more of the above has been influenced by cost and availability and over the years, results have been variable. Since future supplies of any of the above materials are likely to continue to be erratic, every effort should be made to develop self-reliance for this component in seedling production. Some points that should be borne in mind in developing capability are as follows:

1. Material should have a high fibre content to provide internal water holding capacity (small pores) and yet allow for drainage between particles (large pores). Peat and sphagnum moss and sifted compost are good examples.
2. If the material appears oily when wet or is slick rather than fibrous when rubbed between fingers, it is unsuitable for container grown plants.
3. Sufficient pore space within which air can be trapped as roots need oxygen for development and growth.
4. Ability to mix intimately with soil.
5. Ability to bind or hold applied substances such as fertilisers.
6. Should not be a source of fungi, insect pests or weed seeds.
7. Should be sufficiently decomposed as otherwise there may be ammonia damage to roots and foliage or an initially high demand for nitrogen by microorganisms which will induce a nitrogen deficiency in plants growing in comparatively fresh material.

Compost

If sufficient organic raw material is available, compost with the above characteristics can be produced at the nursery site or at a central location. For this reason and in view of its important role in container seedling production, it is mandatory for each nursery to produce at least some of its own compost needs on a continuing basis. A central shredder close to a source of fibre from which sustained amounts could be transported to nurseries would facilitate this activity. In addition to the organic materials already mentioned, the following raw materials are also suitable for making compost and should be more easily available on a sustained, self-reliant basis:

- Weed growth and culled plants at the nursery
- Thinnings and coppice regrowth from woodlots
- Foliage and branches from trees felled in urban centres
- Grass and other road side shrubs and vines
- Bark residue and sawdust

How to Make Compost

Compost is a very valuable addition to any soil and is particularly useful in nurseries where the available soils are either too sandy or too heavy and of low quality. It helps to build a good soil structure, improves water holding capacity of the soil and provides nutrients for plants. It greatly reduces the need for chemical fertilisers and, when mixed with small amounts, dilutes the fertiliser, making it available in much larger useful quantities. It is therefore an economic way of using available chemical fertilisers and is itself a natural fertiliser, very similar to humus.

Composting is the physical and chemical breakdown of organic material by insects, fungi, and bacteria that digest the material during the decomposition process. The two basic ways in which this is done are described as **anaerobic** and **aerobic**. Anaerobic methods supply minimum oxygen to the micro organisms that digest the organic material and is a much slower process, taking periods of up to 8 months or longer. This is because the system utilises deep pits filled with organic material which are usually covered and left unattended. Under these conditions, it is not unusual for foul smelling gases such as methane and sulphur to develop and pollute the surrounding area.

The aerobic method relies on a supply of oxygen and can be ready for use in two months although a 4 month period is considered average. There are three phases based on temperature during aerobic composting and describes them as follows:

"During the first 24 to 48 hours, the temperature in the heap rises to 40 to 50^0 C, destroying sugars and other easily biodegradable substances. During the second phase, as the temperature rises to 55 to 60^0 C, the initial micro organisms die and others specially adapted to the heat begin to break down the more difficult material like cellulose, a component of woody stems. The temperature should reach a peak of 70^0 C for three days to kill all weed seeds and plant diseases. Keeping the temperature between 55 and 65^0 C for as long as possible is the fastest way to produce compost because this is the phase when the most efficient micro organisms are breaking down the hardest to digest material. Turning the pile to incorporate oxygen and to ensure an even distribution of the materials and maintaining 40 to 60 percent moisture allows for optimal composting efficiency. The final stage in which the temperature remains below 40^0 C is called maturing or curing because the bacteria and fungi that help control plant diseases as well as the larger organisms like earthworms, move in."

The leaves of seedlings planted with unfinished compost usually turn yellow because the plant cannot acquire all the nutrients it needs whilst the immature compost continues to absorb what little nitrogen that might be available. To see whether compost is ready, place two moist handfuls in a plastic bag and seal it leaving it in a dark cool place. After 24 hours, open the bag. If no odour or heat is present, the compost is ready. The compost can then be sifted and large particles returned to the next compost batch.

Suitable Ingredients: Compost can be made from most organic materials like grass, leaves, small branches, rice husks, coir dust, bagasse, pulped coffee residues and a mixture is best. One should avoid using woody branches, hard pods and any material that does not break down easily. Dead or rejected seedlings and weeded material from nurseries often make good compost if there is no danger from disease and insect pests. Adding fertiliser may speed up the process and improve its nutrient content but is not necessary and defeats the purpose of producing a cheap fertiliser substitute. The addition of thin layers of rich top soil and or farm yard manure to the first phase and earthworms in the last phase will also prove useful but is not essential.

A small wood lot planted close to the nursery with species that coppice readily such as *Gliricidia, Erythrina* or *Acacia auriculiformis, Acacia nilotica* can be established to provide a steady cheap supply of raw material as required.

Compost Production System: Several closely related methods and modifications have proved successful but the following 'three bed system' is recommended (see Figure 2):

1. Collect ingredients and store excess material in the open since dried material is better.

2. In a flat area, make three rectangular foundations (3 metres long x 1.5 metres wide or of convenient size) side by side using concrete blocks, bricks, or wooden beams. Allow 1 metre between the rectangles as a path for working.

3. On top of this foundation, construct a flat bed with bamboo, wood or metal poles. If coconut or thatch fronds are used, be sure to leave enough cracks and holes for air to pass through and circulate. Choose a material that will not rot too quickly. If discarded wooden pallets are available they can be used for this purpose or laid directly on the ground avoiding the need for the block or brick foundation.

4. Pile organic matter to about 1 metre high on the two outer beds leaving the middle bed empty. Keep the beds flat, not pointed like a pyramid. If top soil or farmyard manure is to be added, it should be spread thinly on top of the pile and a second layer of organic matter placed above. With or without the added top soil, the pile should become warm in about 14 days.

5. After a month to 6 weeks, carefully move the material from the two piles to the empty centre bed starting from the outside of each pile. This mixes the pile, putting the material from the outside into the inside of the pile. After a week, the compost in the centre pile should become warm again.

6. The two outer beds are ready to be filled again with fresh material as soon as the piles have been moved to the central bed.

7. Always monitor the moisture status of the pile. Add water when dry or build a roof if there is too much rainwater. Keep the pile well aerated and moist at all times. Use a plastic sheet, coconut leaves or other covering over the pile to conserve water if the days are hot.

8. After 3 to 4 months the compost in the middle pile should be moved to a storage area. This allows it to mature and be available for use when required.

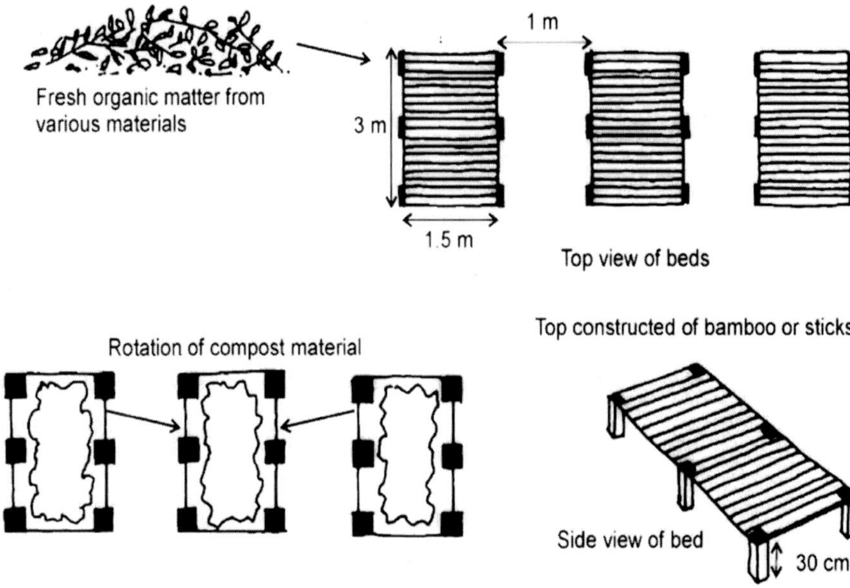

Fig. 2 The 'three bed system' to make compost.

Mixing the Growing Media

With the availability and common usage of such a wide range of organic substances and soil types it is difficult to prescribe an ideal mixing formula. The proportion of soil to compost on a volume to volume basis can range from 1:1 to 3:1. The addition of sand should be unnecessary except with particularly heavy soil.

Assuming that the soil and organic material are on the site free of weeds, weed seed, fungi and insects and is sieved into a uniform and acceptable particle size distribution, the nursery operator must take steps to ensure that quality is maintained.

Some hints for maintaining quality media are:

1. Proper mixing and handling procedures must be followed.
2. The aim is to have a homogeneous mixture with the same measured inputs for each batch. Variability between growing medium batches can result in undesirable differences in plant growth and quality.
3. Mixing is best done on a clean hard surface or concrete slab to avoid contamination and water damage from rain.

4. Once prepared, the mix should be kept under cover if it is to be stored for periods in excess of two or three days.
5. The length of storage determines whether bagged components are stored outdoors or under cover. Most bags begin to deteriorate after a month if left in the open.
6. Media should be available when needed, hence the need to make plans well in advance of the production cycle.

To mix soil and compost

1. Pile components on top of one another, broadcasting any additives over the pile.
2. Take one shovel full at a time and turn it over on top of the pile, working around the edge of the pile.
3. As the material tumbles down the side it gets mixed. Make sure the centre of the pile is moved by gradually moving the location of the pile to one side during the mixture procedure. Mist the pile occasionally to keep the dust down and to make the pile less resistant to water absorbtion.
4. Continue the process until samples from the pile appear to be well mixed.

Media Compaction in Pots

Filling the container with the growing media is an important process because poorly distributed media can negate the beneficial cultural practices of even the best growing media. Under compaction can cause problems but is often ignored. It results in seedlings growing in half or partially filled bags which are small enough as it is and with roots forced to exit containers or become deformed at an early stage.

Over compaction can have several effects on the physical, chemical and biological properties of a growing medium. Although total porosity is naturally less in compacted media, the more important effect is the reduction or elimination of the large pores that control aeration and drainage. Media compaction is difficult to assess because of the small container sizes used and there is no precise technique available for measuring it. Containers that are unusually heavy should be suspect. The medium in properly filled containers should still feel springy to the touch.

The symptoms of over compaction on tree seedling growth are often subtle and difficult to diagnose. They include foliar chlorosis, leaf drop, root browning and eventual death. Because it affects root function, the initial symptoms of root compaction can mimic drought stress, over watering or even mineral nutrient deficiency since roots may malfunction.

Mulching

Mulching means covering the soil surface in bare root beds or individual seedlings with a 5 to 10 mm layer of organic matter. Mulching can provide protection from heavy rain and water splash and it reduces evaporation of soil moisture. In dry localities, the presence of mulch on the surface of transplanted seedlings is a great help in reducing the amount of water required and reduces the tendency for the surface to become muddy or compacted. Like a sponge, mulch can quickly absorb plenty of water which then passes slowly into the soil. In very wet situations, it can be harmful by reducing aeration and increasing the risk of damping off, insect damage and over watering.

11

Nursery Record

Nursery records are valuable documents which should be compiled with due care and attention. They provide particular kinds of information for the nursery man, the forester, the silviculturalist and the economist. The details required vary according to the size of the nursery, but whether the nursery is small or large, good records provide the information necessary for task allocation and for monitoring the efficiency of the day to day operations. The long-term records provide the basis for reviewing the success or otherwise of the nursery methods employed.

The efficiency of a nursery can be judged by its productivity and its costs of production. The nurseryman is required to produce the required number of healthy, vigorous seedlings at the correct time and at reasonable cost. Neither of these can be achieved unless plans are made well in advance and these can only be done effectively by reaching back into the past records to study and compare what has gone on before. It will not be possible to calculate the unit cost of individual operations until the input of labour and materials and the output of work and products are quantified. The cost of plant production is based solely on recurrent labour and materials and need not include capital costs of establishment nor allowance for depreciation of buildings and equipment. These refinements can be developed at Office.

Records of all purchases (seed, chemicals, media, etc.), observation of data (sowing, germination dates and germination per cent, growth, etc.), labor engagement and attendance, sales, pest and disease outbreaks, permanent and temporary stocks (including species wise seedling stocks) and movement register are required to be maintained upto date. Various records of expenditure and income are recorded in different book viz., purchase book, sales book, ledger, cash book, dispatch register, etc. It is advisable to maintain books of accounts for the following reasons: They provide up-to-date nursery business information and guideline for planning. They help to analyze the performance of the nursery activity.

Propagation protocols, growing schedules, and facilities schedules can all be improved each season by keeping good records. It is important to keep records of how to replicate each crop. How long did it take to produce the crop? What materials were purchased? How was the crop fertilized, watered, and managed during each growth phase? What did we learn that will help us next time we grow this species? Propagation protocols should be revised on a seasonal or annual basis, becoming more accurate each year, which then improves the usefulness of the crop and facilities schedules. The best way to improve the accuracy of your protocols is to keep two kinds of written records: 1). The daily log is a journal that notes nursery conditions, activities, and management practices on a daily basis. In other words, what was done? 2). Regular plant growth records describe the development of each crop. In other words, how did the crop respond to particular management practices?

Daily Tasks and Observations

Daily activities include the essentials of keeping the plants alive and healthy such as irrigating, weeding, monitoring, fertilizing, and pest management. Daily observation of the crop is essential for good nursery management and an important way to determine their needs. The manager or a designated "crop monitor" or a nursery man needs to observe the crop every day. This task can be done once daily as a formal practice separate from other tasks. Measurements occasionally will be taken to evaluate the crops progress or to quantify environmental conditions. The person monitoring the crop needs to understand what "normal" is for that crop and for the nursery environment and be highly sensitive to any deviations from that norm. Experience and daily observations can identify potential problems long before they become emergencies.

Observations may include the following:

Appearance: Inspect the crop visually to see if plant size and shoot-to-root ratio are proper for the stage of growth. Look for signs of nutrition or disease problems on the roots and foliage. Inspect closely for insect pests.

Smell: Some problems, such as gray mold, may be discernible to the experienced grower by its odor. Overheating motors, broken fans, composting troubles, and other potential problems can also be detected by the sense of smell.

Noises: Listen for unusual sounds in the nursery such as an engine running unsteadily or water running when it should not be.

Feel: Pay attention to the temperature and humidity in the nursery. Feel the growing medium to determine if it is at the proper moisture level.

Although one designated crop monitor is responsible for this task, all staff need to understand that observation and being alert and aware are key to precluding problems. The manager should welcome and encourage staff to share their observations and contribute to the daily log; this practice builds observation skills and greater crop awareness.

Nursery Calendar

A nursery calendar is an essential tool in nursery planning such as sowing dates and the purchase of supplies and materials. The date for sowing seeds can be calculated by counting backwards from the anticipated date of planting, taking into consideration the number of days needed for germination and further seedling development until the right stage for planting (Table 1). Different species have different requirements for the planting out period (before or during the rains). The time in the nursery also depends on the site on which the seedlings are to be planted. Seedlings for drier sites may need to be larger and need more time in the nursery. Customers might need to be reminded of this when they order plant material to meet certain deadlines. It is also worth anticipating problems with poor germination and/or damping-off to allow time to sow a second time. Once a nursery calendar has been developed, it will help greatly in making decisions about the need for extra labour and requisition of supplies. Consider the likely delays in procuring and shipment of goods, especially when ordering from abroad. Place orders early enough to allow timely arrival.

Table 1. Nursery Calendar used to determine best date for sowing different species

Event	Species		
	x	y	z
Days needed fromtrans-planting to planting out			
Days needed fromgermin-ation to transplanting			
Days needed from sowing togermination			
Safety margin in case of poorgermination or damping off			
Total days needed			
Sowing date from	Determined by counting back anticipated dateof planting in the field		

Daily Log

A daily log is a key record that provides a history of nursery management, problem-solving, and crop development. A daily log (table 2) includes a blank daily log form that a small nursery might use. Make it a habit to at least got down something each day. Large nurseries may keep more complicated daily records and may have separate logs for irrigation, fertilization, and the like. What is recorded in the daily log about management practices, environmental conditions, and general crop performance will become an invaluable resource for many years to come.

Table 2. A sample of daily log

Date
Environmental conditions in growing areas (light, temperature, and humidity)
Sunrise/Sunset times
Moon phase
Other weather notes (cloud cover, rain and so on)
What water did plants receive? (irrigation type and frequency or precipitation)
Today's activities (note how many person-

hours per activity) (fertilization, pest management, transplanting, packing and shipping, making potting media, moving crops from one structure to another, treating or sowing seeds, and so on)

Growth phase status (make notes when a crop moves from one phase to another)

Purchases (what supplies or equipment were purchased and their cost)

Orders (what plant materials were delivered and payments made)

General crop/nursery observations

Questions or concerns

Plant Development Record

Keeping a simple plant development record &/or register for each crop is a great way to build a foundation for accurate, site- and species-specific protocols. Some growers choose to record a large amount of detail in their daily log and then later summarize information about each crop into a plant development record. Others jot notes into a plant development record for each crop on a regular basis; weekly, monthly, or when the crop is entering a new stage in its development. A plant development record notes what is happening with a crop of plants from crop initiation through delivery. At a minimum, you can simply put a couple of fresh sheets of paper in a three-ring binder, make a tab with the species name and date for that crop, and jot down notes on a regular basis as the crop progresses. One way to make it easier to keep track of this valuable information is to use a form such as the one provided in table 3 or by entering information into a computer.

Keep a separate plant development record for each crop grown, even if it is just a small trial of a few plants or even if the species has been grown before. Any intended improvements, trials, or experiments done with a species can also be recorded in a plant development record. When these records are reviewed, the information enables nursery managers to determine if intended improvements actually had a positive effect on plant health and growth compared with what was normally done.

Nursery Inventory

A well-kept and up-to-date nursery inventory helps to assess whether the nursery is operating as planned, and whether demands are being met. Your inventory should list all plants currently in the nursery by bed or frame number, and details of delivery of seedlings, including the site, name of owner and site conditions. It can be an important tool to record feedback from the planting sites and can then help to determine whether seedlings have the right quality for the sites on which they are planted.

Table 3. A sample of plant development record form

Species Name
Propagule source
Date of propagule collection
Establishment
Type and length of propagule treatment (for example, scarified, stratified)
Date of propagule establishment
Growing media and tray or container type used
Germination notes (including date begins and ends, percent germination)
Cutting notes (for example, special conditions, hormone treatments)
Date transplanted (if not direct sown)
Container type and growing media for transplanting
Microorganisms used?
Misting/irrigation (type and frequency)
Fertilization type, rate, and frequency, if any
Environmental conditions for crop (light, temperature, humidity)
Horticultural treatments (for example, cultivation practices)
Date establishment phase completed
Notes (resowing or thinning activities, problems or challenges)

Rapid growth

Time after sowing and sticking to enter rapid growth phase

Plant size at start of phase (height)

Container type and growing medium

Irrigation (type and frequency; for example, daily, every other day, etc.)

Fertilization type, rate, and frequency

Environmental conditions for crop (light, temperature, and humidity)

Horticultural treatments (for example, spacing, cultivation practices)

Date of rapid growth phase completed

Notes (development, vigor and health, challenges or problems)

Hardening

Plant size at start of phase (height and stem diameter)

Irrigation (type and frequency; for example, daily, every other day, etc.)

Fertilization type, rate, and frequency, if any

Environmental conditions for crop (light, temperature, and humidity)

Horticultural treatments (for example, spacing, cultivation practices)

Plant size at end of phase (height and stem diameter)

Date hardening phase completed

Date of plants delivered

Notes (vigor and health, challenges or problems)

Other notes

Notes on performance of crop after outplanting

Placement of Labels

It is useful to make seed and plant labels for each transplant unit of container grown or bare root seedlings. The method often seen in local nurseries is to paint detailed information on rectangular bits of tin or metal which are fixed to pieces of wood in each bed or batch of seedlings giving the nursery the semblance of an untidy cemetery. One alternative is to fold the identification by rolling it so that it can fit into the hollow joint of pencil size bamboo which can fit into another joint so that both ends are closed. The action is similar to that of replacing the cap of a ball point pen. The bamboo with the seed data is kept in the first row of a particular batch of seed or transplants so that it can easily be consulted on the spot when necessary. Labelling transplant beds in this way should not replace the nursery records kept in the office.

Record of Experiments

An up-to-date record of past and ongoing nursery experiments is advisable. Simple experiments testing new potting mixtures, watering regimes, seed pre-treatment's etc. should be part of normal nursery management and, without accurate records of these, valuable information is likely to get lost.

As the nursery grows in size and complexity, entering records into a computer will make the information easier to track. Every day creates small amounts of vital information that will contribute immeasurably to improving nursery management and productivity over time. This information is invaluable for many aspects of nursery management, including the following:

- Budgeting funds.
- Estimating schedules to produce future crops.
- Determining what labour-saving equipment might give the most benefit for the cost.
- Analyzing nursery expenses.
- Improving profits or production.
- Replicating successful crops.

❑❑❑

12

Nursery Business

Starting a nursery business is an exciting and complex process and it can be personally and financially rewarding – if you take the time and make the effort to learn the business and develop a closely reasoned and comprehensive business plan. Success in the nursery business isn't just growing the plants. You must also let everyone know about the business and how your product is superior to the already available products in the trade. Before you can start off, there is some information you should keep in mind about running a nursery business.

The nursery business involves the production and marketing of various plants including trees, shrubs, grasses, perennial and annual flowers, and fruit trees. A landscaping service, garden center, or sod farm may also be associated with a nursery enterprise. A successful nursery operator must be knowledgeable about all phases of plant production and be willing to work long, hard days. Good marketing and management skills are essential. A passion for ornamental plants and an entrepreneurial spirit add greatly to the chances for success.

Learning outcomes

The participants should be able to:

- Define marketing management and its basic concepts.
- Describe the micro and macro environment of the product/enterprise.
- Identify the business threats and opportunities in the market.
- Discuss tools that will aid the sellers in identifying enterprise position in the industry.

- Explain the marketing mix and its implications to marketing techniques and strategies.
- Prepare a simple marketing plan.

Whatever the reasons for starting the nursery business, there are a number of things one should know and resources that should be available. First, success in the nursery business requires expertise in all phases of plant production, from propagation to post-harvest handling.

To be successful, one needs to be conversant with handling both employees and finances marketing and sales. Finally, all business ventures including the nursery business require a capital investment to cover the cost of land, equipment, buildings, supplies, labour and stock, in this case the initial purchase of plants for propagation and growing to salable size. If experienced and having strong knowledge of plants and plant culture, purchasing an existing business or starting a new one are logical options. If case of limited or no experience, this can be gained by working for a reputable grower. Many well established nurseries are always looking for good help as full-time, part-time or even seasonal employees. Deciding to start a nursery is only the first of many decisions to be taken. Objective decisions on all the factors affecting the operation of a business have to be made and a business plan developed.

Starting the business

Starting a nursery business is not all about plants. The focus of the business is one of the most important decisions in the development stage of a nursery. Before the business can have a name, be incorporated, have owners, land, equipment and structures, the following questions need to be answered. What product will this nursery sell? Is there a market demand for the product? Who are the customers that will buy the product? How will be product be distributed and how will it be marketed? After those questions are answered, then the structure of the business can be developed, the type of ownership can be determined, and prospective land can be considered for purchase or development.

Studying current and future market demands and trends, plant material for sale by nurseries and identifying possible niche markets will aid in focusing the future business. After assessing demand and supply, the type of production and the plants to be produced has to be decided upon. Competitors need to be identified and a feasibility study conducted to determine if the demand is sufficient to support another business in the targeted area. A clear purpose with well defined focus for the nursery is crucial to develop a solid business plan, obtain financing, suitable land, and appropriate permits and licenses. Changing production type or plants halfway through developing a nursery is not advisable and can have disastrous effects. Water rights, building permits, zoning restrictions, and loan payback schedules may all be affected if different equipment, supplies or structures are needed or if the production schedule and time to marketing will change.

A business plan should cover the following information:

- A description of the business: name, location
- Business objectives and activities with time frames. The objectives must be SMART Current performance analysis
- Marketing plan
- Production plan
- Financial plan/needs/projections and balance sheet.

Developing a business plan for the nursery

A nursery business plan will help to determine the type, market, location, site, design and size of operation. In addition, one will establish specific details on equipment and employee requirements, operational costs, estimated pricing schemes, projected return on investment and asset availability.

Types of nursery businesses

Many nursery businesses have diversified to provide a variety of products and services to ensure rapid return on their investment. There are three basic nursery formats:

- Landscape nurseries which grow plants for retail sales and for their in-house landscape service.
- Retail nurseries which include small nurseries and garden centres that grow plants on limited acreage for strictly retail sales to the homeowner and to landscape contractors.

- Wholesale nurseries which include contract propagators, contract growers, growers and distributors (re-wholesalers) of nursery stock for wholesale to other nurserymen, landscape contractors and retail outlets.

In deciding on the type of nursery business, one begins to define the type and form (rooted cuttings, seedlings, liners, container grown, field-grown seedlings, or a combination) of nursery stock to be grown. Many of these nurseries operate on less than a quarter acre of land.

Economic Factors

Financial management is a critical aspect of a nursery business and requires expertise in accounting and legal knowledge. Establishing a nursery requires start-up funding to purchase expertise in the planning process, land, permits, equipment, structures, supplies, and labor. Developing plant production and a customer base simultaneously is critical for the success of a nursery.

Sources of money for the establishment of a nursery are usually personal savings of one or several owners and borrowed funds. The amount owners invest in the nursery indicates the commitment of the individual to the business and may determine the amount of credit from financing institutions. A number of options exist for obtaining borrowed funds. Limited partnerships may be developed in which a number of creditors put up the capital needed for the nursery, and they share in the business decisions and the profits. A nursery may engage in contract growing, a procedure by which a nursery agrees to grow a particular crop at a specified price for another firm and in return receives the financing necessary to produce such a crop. Such an arrangement can be mutually beneficial. The borrower has a guaranteed market and the lender has the benefit of purchasing a product at a fixed price. To obtain a loan from a financial institution, the proposed nursery must provide detailed financial data audited by a certified public accountant, information on the objectives and management of the proposed nursery business, and the business organization.

The long time period between investment and first payback is one of the financially daunting aspects of nursery development. Crop cycles of woody plants often span several years. The lack of market information and business profiles may present obstacles in obtaining a loan. One possibility to tie over the first years of no income is to grow fast-maturing crops such as annual bedding plants or other herbaceous crops that can be produced in a few weeks or months and provide cash flow until the target crop is ready for sale.

When establishing a new nursery, many costs need to be estimated. Validation of these estimates by a nursery business in the vicinity or by an experienced grower can be helpful to budget nursery development and production costs accurately. The lending institution generally requires a plan for the payback of the loan. If the nursery land, equipment, buildings, and plants are used as collateral, then these assets are assigned an estimated forced liquidation value. This value is usually quite low. The amount of a loan and specific terms should be negotiated with several institutions, preferably those experienced in agricultural loans, to find the most favourable financing for the new business.

Economics will influence the selection and development of a nursery site. The major factors that will determine success or failure of a nursery business are production costs, plant quality, competition, and available capital. Competition can influence price and in some cases plant quality. Capital availability until plants reach market size and are sold will require several years and is vital for the survival of a new nursery.

Marketing

Nursery crops are marketed in several different ways:

- Retailers market directly to the end consumer, typically homeowners. This is most commonly done either through retail nurseries, which produce some or all of their own plant material, or garden centers, which purchase their inventory from a wholesale nursery. These businesses must be conveniently located for consumer access, ideally near urban or high-traffic areas. Retail nurseries additionally require adequate space and facilities for production, either on-site or at a nearby location.
- Mail order nurseries also sell directly to the end consumer, but their plants are shipped directly to the customer rather than sold at a retail outlet. This is a great option for nurseries that produce specialty plants and whose customers are plant enthusiasts located across the country or globe. The vast majority of mail-order nurseries sell either bare root or small container grown

plants (1-gallon containers or smaller) due to high shipping costs and difficulties in packaging, but larger plants can also be sold by mail-order nurseries if they are highly valuable.

- Wholesalers produce plants that are typically sold in large batches at significantly lower prices to landscapers, retailers, or other nurseries that grow and resell the material at a larger size. Wholesale production is most efficient and profitable when a limited number of plants are grown in large numbers.
- Re-wholesalers purchase large orders of various plants from wholesale producers and resell the plants to landscapers requiring diverse but smaller orders.
- Landscape nurseries produce plants for their own in-house landscaping service, but may have a retail outlet as well.

Marketing Your Nursery

There have been many businesses over the years that have had great products that people would love to consume, yet those businesses failed miserably. Why? No marketing. To be successful in the nursery business isn't just growing the plants, you must also let everyone know about the business and how your product is superior to the already available products in the trade. First and foremost, every business must have a plan to market the product. A marketing plan involves two simple questions: (1) What or who is your market? (2) How will you identify your product(s) to the market?

There are many ways to approach these questions with essentially the same results. A common method in developing a marketing plan is to complete a market analysis. This analysis will help you identify your target market. You must ask yourself, "What is the potential of sales for your product (plants) in your area? Where are the customers located? Will you be selling your nursery stock locally, regionally, nationally?"

Evaluation of enterprises (SWOT analysis)

Activities Participants should be able to list problems they encountered. An exercise to identify their strengths, weaknesses, opportunities and threats should be undertaken.

STRENGTHS	WEAKNESSES
Land and labour	Lack of technical support
Quality of products	Lack of knowledge
Existing customers	Lack of good transport/
Availability of the required	communication
equipment	Lack of support/interest from
the local community	
Experience in growing trees	Lack of water resources
Experienced staff with a broad	Lack of financesü
range of skills and experience	Uneven sale
OPPORTUNITIES	**THREATS**
Availability of labour	Unpredictability of demand
Availability of seedlings	
Market for seedlings	Unpredictable climate
Opening up new markets	Natural disasters e.g.,
in other communities	drought, floods
Join a tree nursery association	Disease outbreaks
in the community to gain more	Financial crisis
knowledge/skills	Policy changes
Deforestation	Increased competition
	Inability to sustain long-term
	investments to enable self-
	sufficiency
	Isolated site leading to the risk of
	vandalism and security problems

Competition

Knowing the competition of your business is essential in developing a marketing strategy. Competition is defined as any other product or brand that a customer may buy instead of the one that is being sold. Competition in the nursery industry may come not only from other tree and shrub producers in the field but from container production, bedding plants and perennial production as well. Understanding what other growers are producing for the market will help to develop a better position for your products and/or services. By developing a competitive analysis one can find out the kinds of products that will sell in the industry, along with pricing strategies and promotions that are necessary for keeping the customer interested in what you do.

After the competition is analyzed and identified, you must find a way to set yourself aside from the rest. In marketing terms, you must find your niche. A niche is defined as a specific situation that is suited to one's character or abilities. In the corporate world, to get the edge on your competition one must have a product or service that others perceive as different from everything else offered, while also fulfilling the wants and needs of consumers better than any other product or service. In the nursery industry there is no real difference. In order to successfully develop a niche, one must determine what it is about your product that makes them better than everyone else's. Is it quality, selection, lowest price, fastest service; what defines your nursery?

For example, a wholesale nursery grower offers basically the same selection you would find at any other nursery, with the exception of one group of plants; Japanese maples. His Japanese maple production began as a hobby until one of his regular customers brought to his attention that the cultivars he had where next to impossible to find locally, that they were available through mail order only. This grower had the means to produce many more Japanese maples, and began to specialize in this particular species. Offering such a product that was currently unavailable locally increased the sales of his other stock simply by getting customers in the door. People began inquiring about his Japanese maples and saw that this was only a small part of his business, and that he grew other quality plants as well. By developing a niche, sales throughout the nursery increased.

What product mix?

After evaluating the environment and identifying the opportunities for the business one can now start strategizing by identifying the target market and delivering the product needed. Marketing of any business can be broken down by using the 4 Ps – Product, Price, Promotion and Placement.

Product

One of the most frequently asked questions is: "What should be grown?" This can be answered through a market survey.

Product development: This will involve:

- Idea generation

- Idea screening: what will it cost to produce? Is it a viable business? Is there demand for the product?
- Concept development/testing: what exactly do consumers want?
- Business analysis: what is the expected production for sustainability?
- Market testing: typical usage of a situation
- Commercialization: product launching and mass production.

Price

You have the product, and know it's time to sell it. What should the price be? First and foremost, the price should be set high enough to generate a profit for the business. One must also make the price attractive to the consumer, getting them excited that a 'sale' or 'special' is taking place. In determining the price, several questions must be asked.

- What are customers willing to pay?
- Are all costs covered; where do you break even?
- What does the competition charge?
- How much profit do you want to generate?

In determining your price it is important to use the cost of production as the base. To determine this, divide the production cost into **variable** and **fixed** costs. Variable costs are **out-of-pocket**, such as labor, materials, and advertising. Fixed costs are those that you will incur **whether you sell anything or not**, such as equipment, utilities, and taxes. Listed below are formulas that are used throughout business in determining your production cost. If after using these models you find that the selling price is more than that of your competitors, you may want to find ways to lessen costs of production or accept less profit without affecting your overall quality of the product.

Formula 1

Materials + overhead + labor (production time x hourly wage) divided by number of units = selling price per unit

Formula 2

Materials + overhead + labor + profit divided by 1 unit = selling price per unit

You must be aware of constant changes in the market such as over saturation and availability that will affect the bottom line.

Promotion

Promotion strategy is everything that a person does to the customer to encourage them to purchase the product. This includes not only advertising but public relations and personal contact. Just as a nutrition company doesn't sell vitamins so much as it sells the benefits of a healthy body; a nursery doesn't necessarily sell only the plant but the aesthetic beauty it will provide in the landscape. You need to decide what the product will do for its consumers. The concept of your promotion strategies are to capture the attention of the public and get them to buy the product.

Placement

Now that the product is ready for sale, priced, and consumers know that it exists, how do they get it? This is known as placement of the product, or getting the goods to the customer when and where they want it (distribution). Distributing the product can represent from 10-50% of the final price of the product in shipping and freight situations.

Often businesses such as nurseries do not have the resources within to individually deliver the product to the consumer. A nursery must rely on the channels of distribution that are established that wholesalers, retailers, distributors, brokers, and cooperatives make up. A nursery can produce the best quality plants around, have a reasonable yet appealing price, and everyone may know about it; but if the consumer has no access it is a formula for failure. On time deliveries and prompt services build the name of a business, promoting word-of-mouth advertising (which is the best kind, FREE!).

Marketing in the nursery industry is just as important as marketing among any other business. One cannot simply rely on people to just 'know' that you are a good grower; you have to make it known. With the proper marketing strategies it is possible to succeed in this business. Remember, growing the plants right and maintaining them is only the first step; people can't buy it if they don't know about it.

Other Ps in marketing

People: the workers must have adequate information on the products and customers.

Process: channel to deliver the product and service.

Physical evidence: portray products in shows, field days and demos.

Packaging: what packaging is required; small or big? Ensure attractive packaging.

A Useful Checklist

The following are suggestions that can help you plan ahead and make you more competitive:

- Calculate exactly how much it costs to produce a given plant. Do not use figures from other nurseries; calculate your own. Be sure to include all costs.
- Selling a product without including the cost of overhead, interest, insurance, etc., will likely prove fatal.
- Determine your profit margin. No business can survive for long without making a profit.
- You may be surprised to discover that producing certain items is not profitable under your management system. You should produce only plants that are profitable.
- Understand both the history and current state of the nursery industry in your area. This will help you benefit from the experience of others.
- Understand the growth potential and the present and projected trends in your industry.
- Understand your direct, indirect and future competitors. It is important to analyze your competitors in terms of product uniqueness, price and quality, marketing strategies, strength, weaknesses and their market share.

- Understand your market segments. Each segment will define the marketing of nursery products differently. The retail garden center, landscape architect-designer, landscape contractor, and wholesale salesman will each require a somewhat different marketing program.
- Understand your customer profiles and find out what pleases them.
- Understand and plan your marketing strategies. How do you plan to promote your products? Consider word-of mouth referrals, personal selling, advertising, publicity, customer service, trade shows, corporate sponsorship, and on-line marketing. Each has advantages and disadvantages. Determine which one or which combination of them can best serve you.

❑❑❑

13

Nursery Law and Regulation

Nursery Registration ACT

For producing certified nursery stock, the fruit nurseries shall have to maintain a scion block, a seed block, and stock bed and shall have to use the propagation material only from these sources. The scion block means a planting of registered trees which serves as a source of scion wood for the propagation of certified nursery stock. The seed block means a planting of registered seed trees which serves as a source of seed for producing seedling rootstock for use in propagation of certified nursery stock. The stool bed is a mean of planting of self rooted registered mother trees/ mother stools which serves the specific purpose of producing vegetatively propagated (clonal) rootstock used in the propagation of certified nursery stock.

Supply of poor quality nursery plants is one of the major causes for low productivity of horticultural, plantation and other crops in our country. Government of India had constituted model nursery regulation act long back in 1954. On this analogy different state governments have also enacted their own state nursery registration acts depending upon the available crops and local requirements to ensure availability of genuine planting material for farmers. A nursery should be registered for the purpose of providing quality seed, rootstock and scion sources. A nursery stock should be inspected, tested for genuinity to type, freedom from viruses, insect-pests and diseases.

Licensing requirements of nursery

1. Any person/ person(s) engaged directly with the distribution of plants, plant products, plant material, nursery stock is required to inform the department, the existence of progeny plants and their operations. He/she should obtain a nursery industry license prior to initiating such business.

2. The aforesaid firm or individual(s) are required to renew their nursery industry licenses, each year or every third year as per local nursery act in order to continue the business.

Inspection of nurseries and nursery stock certification

- After receiving the nursery industry license application, the department or its representative shall examine or inspect all plants, plant material, or nursery stock located or grown on the business location or any other applicable location.
- The department shall conduct inspection yearly or at any time as it deems the best, with or without notice.
- After successful completion of an inspection, the department shall issue a nursery stock certificate to the licensed nursery.

Responsibility of the applicant

The applicant nursery man shall be responsible, subjected to approval by the department, for the selection of the locations and the proper maintenance of registered plantings being grown. The nursery man should maintain register regarding time of different operations/ applications, nursery stock of different varieties of a crop, in a manner approved by the department.

Location of planting: The planting location shall be subject to approval by the department and shall be in an area having minimum risks for spread of infection, pests by drainage, flooding, irrigation or other means. The fruit tree nursery stock being grown shall be on rootstocks approved by the department and shall originate from the registered seed trees or from registered stool beds. Certified nursery stock should be produced on soils that have received a pre-plant nematicide or that were sampled before planting and found to be free from virus, vector and nematodes. Nursery stock meeting the requirements, when sold, shall have the variety; inter stock and rootstock designated, where applicable.

Refusal, suspension or cancellation

Registration or certification may be refused, suspended or cancelled for any plant in part or all of a planting stock under any of the following conditions:

- If nursery is not working as per the standards prescribed by the rules and regulations of state government or Government of India.
- If the plants are found to be virus infected or there are off-type plants in the nursery.
- If a registered mother tree is found virus-infected after periodic indexing.
- If the pest control requirements are not followed properly upon inspection.
- If the plants have not been properly maintained.
- If the registration number is misused.

Civil penalties

The civil penalty for violation of any section of act not already stated will be the assessment of a civil penalty, the confiscation or destruction of any and all plants, plant materials or nursery stock found on the premises or contained in the shipment in question and / or suspension or revocation of the current nursery industry license or any future operation privileges granted under the act.

Guidelines for Recognition of Horticulture Nursery

Availability of true to type, quality planting material is crucial for success of commercial horticulture. This necessitates having a network of Horticulture Nurseries which conform to Model Nursery Standards in terms of Infrastructure, Quality of Seed and Planting Materials and adoption of Nursery Management Practices.

State Governments ensure supply of quality planting materials for fruit crops by enactment of Fruit Plant Nursery (Regulation) Act and enforcement of its provisions through licensing of horticulture nurseries. However, all the States have not enacted Horticulture Nursery Acts. National Horticulture Mission has taken initiative to specify the Nursery Standards by specifying Infrastructure required for setting up of Model Horticulture Nursery etc.

A recognized Model Horticulture Nursery should function as a reliable source of supply of quality planting material for horticulture crops. The recognition shall be accorded to the nurseries for:

a) Production of quality planting material of one or more specified crops by adopting Good Nursery management Practices.

b) Nursery Premise only where sale of specified quality planting material of recognized source are being carried out by creating necessary infrastructure facilities and proper record keeping.

Procedure for nursery recognition and renewal of recognition

- The Nursery seeking NHB Recognition shall apply in Form- I to NHB HQ at Gurgaon. Each such application shall be accompanied by lay out of Nursery showing location of infrastructure components and land utilization plan, details of technically qualified staff in the nursery, major farm machineries and operational manual prepared by nursery for selection and maintenance of mother plants, process followed for production of planting material and management of inventory of planting material. The Nursery shall submit details regarding source of Mother Plants used for propagation of Horticulture plants in prescribed form (FORM- 1) and shall also maintain a register for sale of horticulture plants.
- The application form duly completed along with necessary enclosures shall be submitted to NHB, Head Office at Gurgaon.
- Each application will be considered by NHB based on criteria specified for recognition of nursery.
- The recognition of nursery by NHB shall generally be considered product wise/aspect wise as required.
- On receipt of the application for recognition, assessment would be done by agency identified by NHB with the help of a Technical Committee.
- Agency shall submit its assessment report to NHB along with recommendations.
- On the basis of the assessment report NHB shall decide whether to register the nursery with or without up-gradation.
- The decision of NHB shall be conveyed to the concerned nursery. In case of recognition, the period of validity of recognition with other terms and condition shall be indicated. In case of any deficiency, the nursery shall be given time frame for compliance;

failing which, the application shall be rejected. In case of requirement, additional assessment visits may be undertaken.

Criteria for nursery assessment

The criteria have been aligned with Infrastructural requirement of Modal Nursery and Product specific technical requirement and adoption of Good Nursery Practices for propagation of good quality planting material as detailed below:

a) Product specific criteria to asses for capability and competence of nursery follow technical programme for specific crop as laid down in "Handbook of Seed and Planting Material Testing manual for Horticulture crops" prepared by ICAR.

b) Prescribed Nursery Management Practices and Adoption of Model layout plan.

Assessment Criteria : A system of Graded Certification. It is essentially different from licensing of Horticulture Nurseries under provision of some Act or administrative orders. It is based on continuous evaluation of source of parent material, propagation in disease free condition by adoption of technically prescribed method, adoption of Good Nursery Management Practices, Reliable record keeping and training of staff. Each parameter will be critically examined by assessment team as per laid down criteria. Following grading shall be provided:

Outstanding : * * * * *

Excellent : * * * *

Very Good : * * *

Good : * *

Satisfactory : *

Procedure for application to higher grade

- On receipt of an application for recognition, to higher grade from existing lower grade, the application will be considered to register the nursery, for fresh assesment.
- On the basis of the report submitted and the recommendations of the pre assessment of the Committee, the Competent Authority

will decide on the need to proceed with processing the case of the nursery for up gradation and subsequent recognition or otherwise.

Renewal of recognition

The Nursery seeking renewal of recognition shall apply in prescribed form. The renewal of recognition shall be done based on the satisfactory performance reported as per the surveillance/periodic inspection carried out during the validity period and fresh assessment if felt necessary by NHB.

Processing fee

- Processing Fee of Rs. 5,000/- (non-refundable) shall be submitted in the form of demand draft payable to NHB, Gurgaon along with the application for recognition.
- In case of renewal of recognition, processing fee of Rs. 3000/- (Non-refundable) shall submitted in the form of demand draft payable to NHB, Gurgaon along with the application for renewal
- Government/SAUs/ICAR nurseries shall be exempted from Processing Fee.

Assessment and assessment committee

NHB shall nominate one Agency to coordinate the process who will take in to consideration assessment report from a Technical Committee, hereinafter referred to as Assessment Committee which may have Representatives from:

i) State Agriculture University in the State concerned

ii) Apex Horticulture Institute, NRC or Regional Station of ICAR

iii) Nursery Men Association

iv) National Horticulture Board

v) State Directorate of Horticulture/Agriculture

The members in the Assessment Committee shall be nominated by NHB in consultation with agencies concerned. Assessment Committee shall conduct pre assessment, final assessment and periodic /surveillance of the nurseries. The committee shall submit report and their recommendations to Agency designated by NHB for this purpose.

Validity period of recognition

- In case of approval, initially, recognition shall be granted for a period of two year.
- The effective date of recognition shall be considered from the date of issuance of certificate.

Surveillance/monitoring

Surveillance visit shall be carried out at least once in a year or as required depending on the performance of the nursery. The designated team shall essentially conduct surveillance visit during validity of recognition.

Issue of recognition certificate

- In case, Technical Committee satisfies that the Nursery conforms to the requirements of this recognition, it will recommend for recognition of the nursery. The recognition shall bear an identification number. Recognized Nursery shall have to sign a MOU with NHB for terms & conditions of recognition.
- Any change in the location, layout, design or capacity of the nursery shall be intimated to NHB.
- The date of validity of the nursery recognition shall be specified on the recognition certificate.
- The recognition certificate once issued shall continue to be in force till the date specified in the certificate unless suspended or cancelled at earlier date by Competent Authority. The issuing Authority may institute surprise checks through its officers in order to ensure that the nursery has maintained the standard as required for issue of the recognition certificate.

Refusal/cancellation of nursery recognition

Issue of recognition certificate may be refused or, if issued, may be cancelled or suspended:

I. If the nursery does not conform/fail to perform as per requirements of this scheme.

II. If there are adverse reports from the farmers/users or any other complaints made to NHB by any other entity and upon enquiry duly conducted it is established that the Nursery has breached any of the conditions of recognition.

III. On expiry of the recognition date specified in the Recognition Certificate the recognition ceases to be valid unless renewed.

Appeal against refusal / cancellation of recognition

- Appeal against refusal/cancellation of recognition shall reach to the Managing Director, NHB within 30 days from the receipt of such order.
- In case of cancellation of recognition, the appeal shall accompany the original certificate.
- Managing Director, NHB will consider the application on merit and, in case, found necessary, order re-assessment of the nursery.

Guidelines for Establishment and Management of Spices Nurseries

The Department of Agriculture Cooperatior Farmer's Welfare, Government of India, has entrusted the DASD to put in place a Nursery Recognition Regime for Spices so that a network of recognized Model Nursery is established across the Nation which could function as a reliable source of supply of quality planting materials for spice crops. The spices nursery seeking DASD Recognition shall apply in Form- I attached to this guideline.

Each such application shall be accompanied by following details:

a) Lay out of Nursery showing location of infrastructure components and land utilization plan.

b) Details of technically qualified staff in the nursery.

c) Major farm machineries and equipments in use.

d) Operational Manual of procedures followed for selection and maintenance of mother plants, nursery production techniques and management of planting materials.

e) Details regarding source of Mother Plants used for propagation of each spice crops.

f) Processing fee @ Rs.3,000/- in the form of a Demand Draft in favour of Pay and Accounts Officer, Department of Agriculture Cooperatior Farmer's Welfare payable at Kochi is also to be enclosed along with the application. No processing fee is required for State/Central Government Institutes.

The nurseries should also maintain separate registers for production (stock), sale of seedlings and attendance of staff. Duly filled applications with processing fee in the form of a demand draft should be submitted to the Director, Directorate of Arecanut and Spices Development, Calicut 673 005. Each application will be considered by DASD based on criteria specified for recognition of nursery. The recognition of nursery by DASD shall generally be considered product wise/aspect wise as required.

On receipt of the application for recognition, inspection and assessment will be done by the Technical Committee appointed by DASD. On the basis of this assessment report recognition will be given. The decision of DASD shall be conveyed to the concerned nursery. In case of recognition, the period of validity of recognition with other terms and conditions shall be indicated. In case of any deficiency, the nursery shall be given time frame for compliance; failing which, the application shall be rejected. In case of requirement, additional assessment visits may be undertaken.

Assessment committee

DASD shall nominate one Assessment Committee with a Consultant as head, which may have Representatives from:

i) State Agricultural University in the State concerned

ii) Apex Horticulture Institute, NRC or Regional Station of ICAR

iii) Nursery Men Association

iv) DASD

v) State Directorate of Horticulture/Agriculture.

Assessment Committee shall conduct pre-assessment, final assessment and periodic / surveillance of the nurseries. The committee shall submit report and their recommendations to DASD.

Assessment Criteria, Procedures for application to higher grade, Procedures for application to higher grade and Surveillance / monitoring

Same as on case of horticulture nursery (See the Horticulture Nursery guideline)

Issue of recognition certificate

In case, Technical Committee satisfies that the Nursery conforms to the requirements of this recognition, it will recommend for recognition of the nursery. The recognition shall bear an identification number. The Recognized Nursery shall have to sign a Memorandum of Understanding (MoU) with DASD for terms and conditions of recognition. Any change in the location, layout, design or capacity of the nursery shall be intimated to DASD. The date of validity of the nursery recognition shall be specified on the recognition certificate. The recognition certificate once issued shall continue to be in force till the date specified in the certificate unless suspended or cancelled at earlier date by Competent Authority. The issuing Authority may institute surprise checks through its officers in order to ensure that the nursery has maintained the standard as required for issue of the recognition certificate.

Renewal of recognition

The Nursery seeking renewal of recognition shall apply in prescribed form. The renewal of recognition shall be done based on the satisfactory performance reported as per the surveillance/ periodic inspection carried out during the validity period and fresh assessment if felt necessary by the Directorate (DASD).

Refusal/cancellation of nursery recognition

Issue of recognition certificate may be refused or, if issued, may be cancelled or suspended.

i) If the nursery does not conform/fail to perform as per requirements of this scheme.

ii) If there are adverse reports from the farmers/users or any other complaints made to DASD by any other entity and upon enquiry duly conducted it is established that the Nursery has breached any of the conditions of recognition.

iii) On expiry of the recognition date specified in the Recognition Certificate the recognition ceases to be valid unless renewed.

Appeal against refusal / cancellation of recognition

Appeal against refusal/cancellation of recognition should reach to the Director, DASD within 30 days from the receipt of such order.

In case of cancellation of recognition, the appeal shall accompany the original certificate. Director, DASD will consider the application on merit, and in case, found necessary, order re-assessment of the nursery.

Application Form for Recognition of Horticulture Nursery

FORM-I

1. Name of the Nursery :
2. Constitution : Individual/Partnership, Central/ State Govt. SAUs/Organi -sations under ICAR /CSIR,Co- operative Society, Trust, Company, Otrhers
3. Year of Establishment :
4. Name and Full Address : with Phone/Fax/E-mail :
   a) Owner with Phone/Fax/E-mail
   b) Owner of Nursery Land with Phone/Fax/E-mail
5. Location of the nursery:
   a) State
   b) District
   c) Town or village
   d) Nearest Railway Station
   e) Land Survey No.
6. Total area of the nursery including Mother plants, with survey number

A. DETAILS OF MOTHER TREES:

A1. Own Mother Plants

Area in Ha.	Crop/ Variety	No of Plants	Source of planting materials	Age of plants		Performance record	
				Year	Month	Yes	No

Note: In case of information about source of mother plants is not available, their performance in respect of yield and quality of fruit before being used as mother plant may be given.

A2. Mother Plants are taken Form Other Source

S. No	Crop/ variety	Source of scion	Address	If source is recognised by NHB, please give details	Performance Record Yes No

B. Details of Mother Plants of Root Stock

Area in ha	Crop/variety of root stock plants	Source of mother	No of plants	Age	Performance Record
				Yes	No

- If root stock plants are not his own, please give particulars of the source from which they will be obtained.

S No.	Variety	Source of stock	Address	No of plants	Performance Record	
					Yes	No

C. Number Of Grafted, Budded Or Rooted Cuttings

Give the details of varieties of each horticultural crop, the applicant wants to produce every year.

Crop	Variety	Method of propagation	Details of last two years sales, if any	
			1st year	2nd year

7. Future Action Plan for Fruit Plants/ Spices Production for next three years:

Crop	variety	Method of propagation	No of plants proposed to be propagated during next three years		
			1st year	2nd year	3rd year

8. Details of the facilities for production of disease free planting material

I) Tools & Equipments

S No.	Name of the equipment	Year of Purchase	Number

II) Plants & Machineries including tunnels, shade-house and Poly-House

Type of infras-tructure	Number	Year of erection	Area/ dimension
Tunnel			
Shade-House			
Poly-House			
Others:- Pl. Specify			

9. Source of Irrigation

   a. Ground Water

   b. Surface Water

   c. Micro Irrigation (Drip/Sprinkler)

10. The Sketch plan of the area under nursery and trees showing location of infrastructure components and land utilization

11. Status of implementation of Operation Manual

Flow chart with time scale for production process	Identification of input necessary for each stage of production with specifications	Labeling of each planting material produced	Record keeping

12. Details of Technically qualified staff in the nursery and their training level

S.No	Name	Qualification	Training undergone if any

13. Trade Relationship with recognized Nursery YES/ NO

14. Particulars of application fee paid by Demand Draft in favour of National Horticulture Board payable at Gurgaon or New Delhi and / or Pay and Accounts Officer payable at Kochi, Kerala

Name of the issuing bank	Amount	DD No	Date

15. DECLARATION

a) I hereby declare that information given above is true to the best of my knowledge and belief.

b) I have read the rules and undertaken to fulfill all the conditions prescribed in the rules or as amended from time to time.

Place :

Date :

Signature of the Applicant

❑❑❑

14

Propagation Techniques for Important Tree Species

Timber and Industrial Wood Tree Species

Acacia nilotica: Freshly collected seeds are sown in polybags (after pre-sowing treatment) at 1.5cm depth during Feb-March. Shading is necessary to prevent surface drying. Avoid excessive watering in the first month. Seedlings are fit for planting out in July-August of the same year. Any way one year old seedlings are good for planting in the field.

Ailanthus excelsa: Fresh seed is sown in well worked raised nursery beds, in drills about 20 cm apart, and lightly covered with finely pulverized soil; depth of sowing being not more than 5 mm. Seed being minute is mixed with ash or fine soil to ensure even sowing. Seed beds are irrigated regularly but lightly as the seedlings are liable to damping off. Germination commences in about 10-12 days and taken about 50 days to complete. Young seedlings are susceptible to insect and rat damages. In order to avoid it, porcupine proof fencing of nursery is helpful. Seedlings raised in the containers attain 15-25 cm height in the first growing season, while the subsequent growth is more rapid.

Albizia lebbeck: Seedlings can be raised by sowing in drills 15 cm apart in the nursery beds in March, or in containers. Germination commences in 4 days and is complete in 17 days. Prick out seedlings into polybags when they attain 7.5 cm tall. Seedlings growth is fast and is susceptible to insect attack also. Seedlings reach plantable size in July of the same year.

Azadirachta indica: Depulped, fresh, dried seeds are sown in nursery raised bed, in lines 15- 20 cm apart at 1 cm depth. Nursery beds are watered sparingly; weeded, hoed and protected against frost in winter. Germination starts in about a week and continues for 3 weeks. Plantable size seedlings (20-30 cm tall) are produced in 2-3 months. But the pricked out seedlings are maintained in the nursery for one year to increase its field performance.

***Bamboo* spp.:** Bamboo reproduce themselves naturally from seed fallen on the ground after sporadic or gregarious flowering, also from the rhizome buds, throwing up culms as annual shoots. Bamboo can be propagated in the nursery by direct sowing of seeds, planting nursery raised container or bare rooted seedlings, and by vegetative means. Most common method of vegetative reproduction is by rhizome or offset planting; other methods are layering, nodal cuttings and culm cuttings.

***Dalbergia sissoo*:** Broken pod segments (dewinged pods) or clean seeds are sown about 1.25 cm deep in nursery beds, 0.9 m to 1.2 m wide, flat, or sunken beds during Feb- March. Sowing may be done any time between mid-February and September, but late sowings give lesser growth. Bed needs to be irrigated at least twice in a month. For planting out, 3-4 months old entire plants, when about 60 cm tall are suitable; for making stump 12-16 months old stocks are preferred. Before digging out plants, most of the lower leaves are stripped off to reduce transpiration.

***Dendrocalamus strictus*:** Seed is broadcast in raised nursery beds, 10 m x 1 m, in September - October; seed rate being approximately 1.5 kg per bed. Germination takes 10-20 dates; seedlings when about 3 months old and 4-5 cm tall, are pricked out in polythene bags. Seeds may also be sown direct in polythene bags. Seedlings with 45-60 cm tall are fit for planting out during following July.

***Eucalyptus* spp.:** For raising container plants seed is first sown in raised nursery beds and then pricked out into polythene bags, after germination. About 200g seed is sown per germination bed of 10 m x 1 m size, during September-October or after Feb-March after winter frost. To avoid dense sowing, seed is mixed with fine sand or earth, and given a light covering of fine sand; Chlorpyriphos @ 100g per bed is advised to protect seeds from insect damage. Germinating beds are covered with thatch grass to prevent drying out and minimize bird's damage, or direct impact of water spray of a chance shower of rain. Germination takes place within 5-15 days, and then thatch is removed. Seedlings are pricked out into polythene bags when about 5-10 cm tall, normally 4-6 weeks after germination. These bags are placed in sunken beds and watered regularly to its 30-45 cm height and are fit for planting in 6-8 months. Eucalyptus is also propagated by two leafy cuttings which require mist chamber and other propagation hardware. This ensures true to type propagules and higher productivity.

Gmelina arborea: Seed is sown in drills, as soon as it ripens, in nursery beds; shading is necessary. Germination starts in 2-3 weeks. Seedlings are transplanted in to polybags and maintained up to 6 months for its height growth till > 20 cm.

Tectona grandis: Commonest type in moist area is the temporary nursery, while dry and irrigated nursery is recommended, for dry and drought prone areas. The soil is well dug up, mixed with ashes, in a well-drained site. Standard size of bed is 12 m x 12 m raised bed in moist localities to avoid water logging and flat or slightly sunken bed in dry zone is advised. Pre sowing treated seeds are sown in broadcast or 7.7 x 7.5 cm lines, about one cm deep, between February and May. Seedlings are pricked out in transplant beds during the first rains and one year old seedlings are suitable for the stump preparation. 1.0 to 2.0 cm collar diameter; 3-4 months old seedlings are suitable for planting.

Fuel Wood Tree Species

Acacia auriculiformis: Seeds requires pre-treatment, either 24-48 hours immersion in water at room temperature or boiled water for 24 hrs. Sowing in bed/polythene bag (20 cm x 15 cm size) during March to early April is suggested. Seed is sprinkled with kerosene oil to protect against ants. Germination starts in 30-35 days. A thatch barrier on one side of bed is needed to protect the young seedlings against hot winds. 15-30 cm tall seedlings are ready for planting in the fields.

Acacia senegal: Seeds are sown in polythene bags filled in with a mixture of good loamy soil and FYM in the ratio of 2:1 during June-July. One year old entire seedlings are planted out in the following monsoon. Seedlings should be protected against drought and frost by regular watering.

Casuarina equisetifolia: Nursery soil should be inoculated with soil collected from *Casuarina* plantation, and quinalphos sprinkled around beds to prevent ant damage, seed is sown broadcast @ 10 g/

sq. m size bed, usually in November, and lightly covered. Beds are mulched with straw and watered daily till germination is complete (20-30 days). Seedling are fit for planting when 35-45 cm tall in 5-6 months.

***Pongamia pinnata*:** Seed is dibbled in the nursery beds any time after collection, preferably in the beginning of hot weather at a spacing of 7.5 cm x 15 cm. Mulching of sown beds is helpful. Germination commences after about 10 days and completes in a month. Pricking out improves growth. Seedlings attain a growth of 25-30 cm by the end of first growing season, and of about 60 cm by the beginning of the next rainy season, when they may be planted out entire, with ball of earth, or in the form of stumps.

***Prosopis cineraria*:** Pre-sowing treated (acid scarification for 12-15 minutes) seed is sown in soil filled perforated polythene bags; 2 seeds in each, I cm deep, in June-July. Regular watering is necessary, but excessive watering is harmful. Seedlings are regularly weeded; if two emerge, one of the two seedlings is removed. They reach planting size after one year.

Fodder Tree Species

***Acacia leucophloea*:** Seeds soaked in cold water for 24 hrs are used to sow in nursery. Germination starts after a week and is complete within a month, then they are transplanted to the polybags. The seedlings are ready for planting after six months of nursery period.

***Albizia procera*:** Treated seeds (hot water or cold water soaking for 12 hrs) are sown in nursery beds in drills 7.5 x 7.5 cm during May. Light watering is given for about a week. Germination commences in 3 to 4 days and completes in 3 weeks. Seedlings attain about 13-15 cm height in July of the same year, are fit for planting out, with or without ball of earth, during rains.

***Leucaena leucocephala*:** Seeds require inoculation with *Rhizobium* culture to ensure nodulation and nitrogen fixation. Plants may be raised in nursery beds or containers. Preferably inoculated seeds are sown in beds in March-April, in lines spaced 30 cm apart and seeds 5 cm apart in lines and about 1.2 to 1.5 cm deep. Nursery soil should be neutral to mildly alkaline. Beds are irrigated regularly but lightly. Germination commences in 4-6 days and the seedlings attain plantable size by July-August.

***Moringa oleifera*:** Seed is sown in June, in sunken nursery beds, in lines 20 cm apart, seeds about 2 cm apart in lines and just, about 1 cm deep. Addition of well rotten compost to the nursery bed is beneficial. The required seed rate is 30-35 g/sq. m of bed. Germination starts in about 8-10 days and takes 4 weeks to complete. The seedlings are fit for planting out, and also for making stumps after one year.

***Sesbania grandiflora*:** Seeds are sown in polythene bags after collection in May and watered regularly. Germination starts on the third day of sowing and will complete within a week to ten days. Seedlings grow very rapidly and are fit for planting out in about one to two months.

***Ziziphus* spp:** Treated seed are sown broadcast, or in lines in nursery beds, or directly into containers, in February-March, and covered with soil to a depth of 1.5 to 2.0 cm. Seedlings are pricked out into transplanted beds/containers at the 4-leaf stage, at a spacing of 30 x 15 cm. Plants attains plantable size (>30 cm) in the first or second rains. Cultivated varieties are usually propagated by ring grafting or budding on wild rootstocks, either *in-situ* or in the nursery seedlings.

❑❑❑

15

Nursery Management Tips

1. Potting Mixture (Nursery-Soil)

There is a tendency to collect the topsoil from forest area to fill the polythene bags. This practice is not good as it contributes to soil erosion. The whole success of nursery depends on good soil mixture. It should be light in weight, well decomposed and well drained and free from insects as well as weed seeds.

Always use compost in the nursery. It can easily be prepared by decomposing organic materials (green leaves). It increases the nutrient contents of the potting mixture and enhances the air spaces and water holding capacity of the mixture.

2. Seed Collection and Sowing

- Always collect seeds from plus trees or phcnotypically superior trees.
- A chart should be prepared in each nursery indicating the seed collection period of local tree species along with the location of plus trees.
- Seedlings developed from poor or abnormal trees will never produce good trees.
- After collection seeds should be processed carefully otherwise they may get damaged and loose viability.
- Each species requires different processing after seed collection i.e. seeds with pulp (e.g.-kadam) are to be processed in a different way than the pods (e.g.-mandhani), drupes (e.g.-bonsum and dhuna) and capsules (e.g.-khokan).
- In tropical area most of the seeds have short viability, therefore, sowing is to be done immediately after collection and processing the seeds.

- Sow large seeds 2cm to 3cm deep in the soil.
- Sow small seeds (e.g.-Kadam, Khokan) about 0.5 to 1cm deep or broadcast and cover them with a small amount of fine soil. The beds can also be covered with thatch or plastic sheet till the germination of seeds.

3. Watering

- Keep seedbeds moist not wet.
- Do not water at a fixed time each day. Water when the plants need it.
- Small seedlings require less amount of water.
- Large seedlings require more water more often.
- Seedlings require more water more often on windy days and sunny days.
- Water less often when seedlings are kept in shade.
- The moss and algae growth is an indication of excess watering.
- Always use clean water; dirty water may cause disease problem.
- Over-watering results into weak plants and causes diseases such as root-rot and damping-off of seedlings.

4. Transplanting

- Water mother beds thoroughly before transplanting the seedlings.
- Always use some tools (bamboo-sticks, etc.) to loosen the soil before pulling out seedling from the beds.
- Make a deep and wide hole in the polythene bag or container for transplanting the seedling.
- Hold seedlings at the base of the stem and pull it out gently from the mother beds.
- Never bend the roots and do not force the seedlings to fit into the hole.
- Keep transplanted seedlings under proper shade until they have recovered.

5. Hardening of Seedlings

In a nursery seedlings are kept under ideal condition, therefore, hardening is essential to make them strong enough to tolerate the harsh conditions of field/planting site. It is generally done by reducing the amount of watering about one month before the date of planting.

6. Nursery Hygiene and Disease Control

Most of the micro-organisms, insects and pests that cause diseases in the nursery live in weeds, trash and puddles. Therefore, keeping nursery in neat and clean condition to reduces the chances of attack of common diseases.

All trash, waste, polythene bags and diseased plants should immediately be removed and burnt at a place far from the nursery.

- **Pesticides Recommended** : Diazinon. Melathion, Sevin
- **Fungicides Recommended** : Bavisteen. Benlate, Captan. Dithane M-45
- **Dangerous Pesticides** : Aldicarb. Aldrin, DDT, Lindane, Nuvacron, Heptachlor, etc. These are very dangerous.Never use them.
- **Natural insecticides** : Neem Seeds: Make powder of 500g dried neem seeds and soak overnight in 15 litres tap water. Sieve it twice and spray. Besides this, chilli powder and tobacco leaves powder can also be used. They generally take little longer to repel or kill the insects, therefore, apply them immediately as soon as insect problem is noticed. The main advantage is that they are natural, hence safe and biodegradable.

Remember the following things while using them

- Proper doses should be prepared by carefully reading the label or guidelines.
- Always provide protective clothing, mask, gloves and goggle to the person spraying the pesticide.
- Never mix insecticides and fungicides together in the same sprayer.
- Never eat, drink or smoke while spraying.

- Never use pesticides containers to store other things.
- Extra pesticides should be disposed by digging a hole away from rivers and wells.

Common Errors Found in Nurseries

- Containers not filled properly: After a short period, the growing media settles leaving space at the top, exposing roots. The top of the bag folds inwards and consequently during irrigation, insufficient water reaches the seedling.
- Cylindrical shape of container not maintained: If the plastic bags take on different shapes in the transplant beds, each seedling will not receive the same optimum space for growth and light.
- Container not in upright position: Plastic bags not placed in an upright position is one of the causes of stem curvature.
- Soil or sand used in germination beds not changed after each production cycle: Using the same germination media each year increases the risk of attack from damping off fungi. Preventive methods are more effective than control.
- Sowing seed too deep: This is often done to avoid the washing out or displacement of seed by rain or heavy watering and results in unequal germination or low germination percent.
- Lifting transplant seedlings individually and wrenching them: Transplant seedlings should be lifted carefully in groups with the aid of a wooden knife to avoid destruction of the young tender roots and root hairs.
- Exposure of seedlings to air after lifting: Exposure to air for longer than ten seconds can cause unnecessary wilting and delay normal seedling growth.
- Handling transplant seedlings by root or stem or at root collar: Always hold seedlings by the foliage to avoid physical damage caused by abrasive dirty fingers to delicate roots and stems.
- Bad transplanting: If principal or secondary roots are bent upwards, intertwined or benched, a poor quality seedling will be produced with undesirable roots.
- Leaving air space around the root of the young seedling after transplanting: Roots lack contact with soil and are unable to fulfil

function of water and nutrient uptake resulting in wilting, slow recovery and sometimes death.

- Bad root pruning while transplanting: Nursery workers use their fingers and fingernails to cut back long roots during transplanting. In doing so they squeeze the roots between their fingers, damaging them. Roots should be cut with a sharp knife.
- Uneven transplant container beds: Water collects in depressions causing chlorotic symptoms and stunted growth.
- Same knapsack sprayer used to apply weedicides and fungicides: Large quantities of seedlings have been killed by this practice even with thorough washing.
- Inadequate attention paid to root pruning in transplanted containers before transporting to field: Insufficient time available for the seedling to recover from shock and seedling will wilt at the planting site.
- Hardening off process starting too late or neglected: Do not wait until the last week or two to start this operation as seedlings will not have had sufficient time to adjust to the harsher environment in the field. The rains will also have started and controlling water reduction at this time will be difficult.
- Dispatch and transport of plants not properly done: Often plants are lifted by their stems when loading and unloading vehicles and when transporting to the planting site. Never lift plants by the stem. Boxes or crates should be provided for the purpose.
- Too many trees in the nursery: Tall trees give unnecessary shade, harbour insects and other pests and make it difficult to achieve a good hardening process.
- Bad storage of plants near planting site: Try to avoid leaving plants in the shade without water for long periods after they leave the nursery. Remember that bad handling and poor care after seedlings leave the nursery often destroys in a few days all the skill and care taken by nursery workers during the previous 3 to 4 months.

❑❑❑

16

Glossary

A

Absorption: Movement of a pesticide from the surface into a body of water or of nutrients into a plant (compare with **adsorption).**

Acid soil: Soil having a pH value less than 7.0.

Active ingredient (a. i.): Portion of a pesticide formulation that produces the desired toxic, stimulatory, or repelling effect, expressed as a percentage.

Adjusted sodium adsorption ratio (ASAR): Irrigation-water quality index that is the ratio of deleterious ions (i.e., Na^+, CO_3-, and HCO_3-) to beneficial ions (i.e., Ca^{++} and Mg^{++}). The ASAR value is used to assess the effect of salts on soil permeability (compare with **residual sodium carbonate, RSC**).

Adsorption: Attraction or bonding of ions or compounds, usually temporarily, to the surface of a solid (compare with **absorption**).

Advection frost: Frost occurring when cold air flows from a higher location to a lower one, displacing lighter, less dense air and often pooling to form a frost pocket (compare with **radiation frost, wind frost**).

Aeration (soil): Process by which (1) oxygen diffuses through the soil to the root and (2) carbon dioxide and other gases released from the root diffuse to the soil surface.

Aerobic: Occurring or growing in the presence of oxygen.

Alkaline soil: Soil having a pH value greater than 7.0.

Allelopathy: Production of chemical compounds by one plant which are released into the soil environment and are harmful to other nearby plants.

Amendment: Any substance added to a soil to alter its physical or chemical properties and thereby make that soil more useful for plant production.

Anaerobic: Occurring or growing in the absence of oxygen.

Anion: Ion having a negative charge, e.g., Cl -, NO_3^- (compare with **cation**).

Annual: Plant that completes its entire life cycle from seed germination to seed production and death within a single season (compare with **perennial, biennial**).

Artificial regeneration: Reforestation of a stand by direct seeding or planting (compare with **natural regeneration**).

Available nutrient: Any essential element or compound in the soil that can be readily absorbed and assimilated by plants.

Available water: Portion of soil water that can readily be absorbed by plant roots. Generally considered to be that water held in the soil within a water-potential range of approximately - 1/3 to -15 bars.

B

Band application: Spreading of a chemical or fertilizer to a restricted area (such as in, on, or along a crop row), rather than over an entire field or area (compare with **broadcast application**).

Bar: Unit of pressure equaling 1,000,000 dynes/cm2; typically used to measure plant and soil water potential.

Base saturation percentage: Proportion of the cation exchange capacity (CEC) occupied by cations other than hydrogen or aluminum; expressed as a percentage of the total CEC.

Bed: Elongated strip of soil in which seedlings or transplants are grown.

Bed foot (meter): Area of seedbed 1 lineal foot (or 1 lineal meter) long times the width of the bed.

Biennial: Plant that requires 2 years to complete its entire life cycle, from seed germination to seed production and death. Food is stored in the first season and then used in the second season when flowers and seed are produced (compare with **annual, perennial**).

Biometrics: Use of statistical techniques to evaluate biological problems.

Box pruning: Root-culturing technique that consists of the lateral pruning of roots in a four -sided box shape around a seedling in a seedbed (compare with **lateral pruning**).

Broadcast application: Spreading of a chemical or fertilizer over an entire area or field rather than only on rows, beds, or individual plants (compare with **band application**).

Broadcast seeding: Method of sowing in which seeds are distributed across the seedbed by hand (compare with **drill seeding).**

Broadleaf species: Those plants classified as Dicotyledoneae: characterized by having round and flattened leaves (compare with **narrow leaf species**).

Brushing: Frost-protection technique in which shields are placed near plants to deflect radiation to the plant and soil during the day and reduce heat loss to the sky at night; used primarily for agricultural crops.

Buffer capacity: Ability to resist change in pH. A soil with a high buffer capacity will have stable soil pH.

Bulk density (soil): Mass (weight) of dry soil divided by soil volume, commonly expressed as pounds per cubic foot or grams per cubic centimeter.

C

Caliper: Stem diameter of a seedling, usually measured just above the root collar.

Candle: Terminal shoot of conifers, especially pines, that has elongated but not yet produced needles. Color of bud scales or a waxy secretion gives the terminal shoot the appearance of a white candle.

Capillary water: Water held in the small pores of a soil, usually with attraction forces exceeding the pull of a 60-cm column of water. Most capillary water is available to plants.

Carbon:nitrogen (C:N) ratio: Ratio of carbon content to nitrogen content. The C:N ratio in plant residues is often a convenient predictor of decomposition rates but is not the only determinant.

Catch crops: Crops which are grown and then incorporated into the soil, primarily to capture and retain nutrients on the site but also to increase soil organic matter (compare with **cover crops, green manure crops**).

Cation: Ion having a positive charge, e.g., Ca^{++}, Mg^{++} (compare with **anion**).

Cation exchange capacity (CEC): Sum total of exchangeable cations that a soil can adsorb, expressed in milliequivalents per 100 grams of soil, clay, or organic colloid. Because of the high CEC of organic matter, the buffer capacity of a soil is increased after organic amendments are added, and the soil is better able to hold cationic nutrients.

Chelation: Formation of strong bonds between metals and organic compounds. Those chelates used in fertilizers are soluble and increase and maintain the availability of micronutrients (such as iron, zinc, manganese, and copper) for plant uptake.

Chilling requirement: A certain period (usually expressed in hours) of cold temperatures (0 to 10°C) that seeds or seedlings must experience before growth occurs in periods of warmer temperatures. This adaptive mechanism helps to protect buds from flushing during temporary warm temperature spells in winter or early spring and being subsequently killed when cold temperatures return.

Chilling sum: Cumulative number of hours that a seedling is exposed to a certain range of temperatures, of either soil or air, known to effectively release dormancy for the species of interest (e.g., for Douglas-fir in the Northwest, the temperature range is 0 to 5°C). Chilling sums are used to assess seedling dormancy status and to determine proper lifting dates.

Chiseling: Breaking or loosening soil, without inverting it, witha cultivator or chisel plow, generally below the normal plow depth (compare with **subsoiling, ripping**).

Chlorosis: Yellowing of normally green plant tissue, due to a lack of chlorophyll. Chlorosis can be a symptom of disease, nutrient deficiency, or inadequate light.

Clay: Soil particle less than 0.002 mm in diameter; soil textural class characterized by a predominance of clay particles.

Claypan: Dense, compact layer in the subsoil which has a much higher clay content than the overlying material, resulting from the downward movement of clay or the synthesis of clay in place during soil formation. Claypans, separated from the soil material above by a sharply defined boundary, are typically hard when dry and plastic and sticky when wet. They usually impede water and air movement and growth of plant roots (compare with **hardpan**).

Coefficient of variation (CV): Relative measure of variation in which the standard deviation is expressed as a percentage or fraction of the treatment mean.

Compaction (soil): Increase in bulk density, hence lower porosity, of a soil, due to the rearrangement of soil aggregates from applied loads, pressure, or vibration. The reduction of pore spaces between particles impedes gas and water exchange and also root penetration.

Complete fertilizer: Fertilizer containing all three of the major fertilizer elements—nitrogen (N), phosphorus (P), and potassium (K): the concentration is expressed as the percentage of N but as the percentage of the *oxide* of phosphorus (P_2O_5) and potassium (K_2O).

Compost: Organic residues or a mixture of organic residues and other materials (e.g., sawdust combined with nitrogen fertilizer or sewage sludge) that have been piled and allowed to undergo biological decomposition.

Concretion (soil): Local concentration of a chemical deposit, such as calcium carbonate or iron oxide, in the form of an aggregate or nodule of varying size, shape, hardness, and color.

Confidence Interval: Specified range within which the quantity of interest (e.g., treatment mean) will be contained for a specific level of confidence (e.g., 95%).

Contact herbicide: Herbicide that kills plant tissue by direct contact rather than by translocation or root uptake (compare with **systemic herbicide).**

Control treatment: Zero-level application of a treatment. A control treatment (e.g., no wrenching) is used to judge whether particular treatment levels (e.g., multiple wrenchings) are effective (compare with **standard treatment**).

Cotyledon: First leaf or leaves of the embryo in seed plants. In conifers, the cotyledon stage occurs after the seedling has emerged and until the primary (true) leaves develop.

Cover crops: Crops grown principally to control various forms of erosion but also incorporated into the soil to increase organic matter (compare with **catch crops, green manure crops**).

Cull: Seedling which is not acceptable because it does not meet certain size and quality standards and which is thought to have low survival and growth potential.

Cull factor: Number of seedlings that do not meet shippable standards (e.g., diseased, poor form or size, damaged), expressed as a percentage.

Cultural control: Indirect control measure (as opposed to direct killing) used to prevent pest damage—e.g., use of a previous crop that is antagonistic to the pest.

D

Damping-off: Disease characterized by either seed decay in the soil or seedling wilting and death after germination, usually caused by soil-borne fungi.

Deciduous plants: Plants which shed their leaves at a certain season, usually autumn.

Deductive reasoning: Drawing conclusions or making inferences which logically follow from a well- defined base of biological, chemical, or physical principles (compare with inductive reasoning).

Designed experiment: Detailed investigation in which the experimenter or researcher requires precise and unbiased (1) conclusions and (2) measures of uncertainty associated with those conclusions. Treatments are usually replicated and applied to plots at random (compare with operational trial).

Determinant growth: A 2-year process of shoot growth in which primordia are laid down in buds in the first year but elongate in the second year. Species exhibiting determinant growth may only have one growth flush per year (compare with indeterminant growth).

Dicot (Dicotyledoneae): Larger of the two classes of flowering plants (Angiospermae), distinguished from the smaller class (Monocotyledoneae) by the presence of two seed leaves (cotyledons) in the embryo and by other structural features, e.g., net -veined leaves; includes oak, elm, alder. Certain herbicides (e.g., 2,4-D) are effective against dicots but do not harm monocots.

Disking: Breaking up surface layers of soil with a disk implement, to destroy weeds, prepare the soil for planting, or incorporate a pesticide or fertilizer.

Dormancy: Condition in which a tissue predisposed to elongate does not do so even if environmental conditions are suitable for growth. Dormancy, composed of different phases, is a plant adaptation to survival under stress (e.g., frost, drought).

Drench: Saturation of a soil with pesticide, usually to control root diseases.

Drill seeding: Nursery sowing method in which seeds are planted in rows with a seed-drilling implement (compare with broadcast seeding).

Dysgenic: Causing harmful changes in the genetic makeup of a population.

E

Ectendomycorrhiza(e): Group of mycorrhizae which have both intercellular and intracellular fungal penetrations of root cortical cells. The branching and Hartig net formation are similar to those in ectomycorrhizae; infection within cortical cells is similar to that in endomycorrhizae.

Ectomycorrhiza(e): Group of mycorrhizae in which the fungal hyphae penetrate between the host root cells, often forming a mantle or sheath over the feeder roots. Ectomycorrhizae are common on members of the Pinaceae, Fagaceae, Betulaceae, and Salicaceae.

Electrical conductivity: Reciprocal of electrical resistance, expressed in "mhos" (reverse of "ohms"). A method for expressing salt concentration in soil or water.

Embryo: Young plant developing after the union of male and female gametes. In seed plants, the embryo is contained in the seed.

Emulsifiable concentrate (EC): Liquid pesticide formulation consisting of an active ingredient, a solvent, and an emulsifier that mixes with water to form an emulsion.

Emulsifier: Material which helps to suspend one liquid in another, such as oil in water.

Emulsion: Mixture of two or more immiscible liquids, such as oil and water, in which one is suspended or dispersed in the other in the form of very minute droplets and remains so through the use of an emulsifier.

Endomycorrhiza(e): Group of mycorrhizae in which the hyphal infections of host roots are intracellular. This group is not as common in conifer species as it is in many angiosperms and herbaceous species, although cedars and redwoods have endomycorrhizae.

Evapotranspiration: Sum of the water transpired by vegetation plus that evaporated from the soil.

Experiment: Planned inquiry designed to obtain new facts or to confirm or deny information from previous results, to aid in making recommendations or decisions.

Experimental plot: Smallest physical unit (e.g., specific length of nursery bed) to which a treatment is applied independent of other treatments.

F

Factorial experiment: Experiment in which each level of a given factor is tested across each level of one of more other factors.

Fallow: Allow cultivated land to lie idle during the entire or greater portion of the growing season.

Fertilizer: Any organic or inorganic substance, either of natural or synthetic origin, which is added to the soil to provide elements essential for plant growth.

Field capacity: Soil water content resulting after the free water has been allowed to drain from a saturated soil for 1 to 2 days; expressed as a percentage on a dry-weight basis.

Field efficiency: Amount of area growing trees divided by the amount of area cultivated in a given field: expressed as a percentage. Field efficiency primarily depends on the distance between irrigation lines and the width of tractor paths.

Frost cracking: Type of cold injury caused by freezing and thawing in which exterior and interior portions of the bark expand and contract at different rates, causing the bark to crack.

Frost hardiness: Ability of plant tissue to survive and resist the stress from freezing temperatures without sustaining irreversible physical damage.

Frost heaving: Lifting of the soil surface due to growth of ice crystals in the underlying soil; when this recurs over a period of time, seedlings can be physically lifted out of the ground.

Frost pocket: Area whose topographic features cause cold air to accumulate, increasing frost hazard to seedlings.

Frost smothering: Condition occurring when a saturated soil freezes, allowing little or no oxygen to reach plant roots. If the condition persists over 48 hours, seedling damage or death can result.

Fumigant: Chemical applied as liquid or powder which volatilizes to gases and kills insects, nematodes, fungi, bacteria, seeds, roots, rhizomes, or entire plants. Fumigants are usually applied beneath a tarp, sheet, or other enclosure.

Fumigation: Use of chemicals in gaseous form to destroy pests, usually applied under a cover or shelter. **Fungicide:** Chemical used to kill or inhibit fungi.

Fungistasis: Inhibition of fungal growth, without destroying the fungus, by preventing the germination of conidia and other spore types.

G

Genotype: Genetic constitution of an organism, i.e., the set of genes belonging to an individual. Genotype interacts with environment to produce the phenotype (compare with **phenotype**).

Germination: The beginning of growth of a mature, generally dormant seed.

Germination percent (seed): Percentage of seeds that germinate under standard treatment and after a given time period. This value, considered a principal index of seed quality, is used to calculate seedbed sowing density.

Germinative capacity: Number of seeds in a given sample that actually germinate regardless of time required, expressed as a percentage.

Germinative energy: Number of seeds that have germinated at the time of peak germination, expressed as a percentage.

Gley: Layer of mineral soil developed under conditions of poor drainage, characterized by reduction of iron and other elements, and gray colors and mottles.

Grading: The Process of identifying and subsequently separating various classes of acceptable (shippable) and inferior (cull) stock to improve stock quality is called grading. This operation occurs after lifting and before packing and storing.

Gravitational potential: Component of soil water potential caused by the force of gravity.

Green manure crops: Crops grown primarily as organic amendments for the soil. Green manure crops are incorporated into the soil while green but before seedset, to benefit succeeding crops (compare with **catch crops, cover crops**).

H

Hardening off: Natural process of adaptation by plants to cold or drought. Hardening off may be induced in the nursery by reducing water or by root culturing, thus preparing the seedling for outplanting or transplanting.

Hardpan: Hardened soil layer caused by cementation of soil particles with materials such as silica, sesquioxides, or calcium carbonate. The hardness does not change appreciably with changes in moisture content (compare with **claypan**).

Hartig net: Hyphal network of ectomycorrhizae which penetrates between root cortical cells of the host. This extensive contact between fungus and plant root facilitates the exchange of nutrients and other substances.

Herbicide: Chemical used to kill or inhibit unwanted plants or weeds.

Hormone: Growth-regulating substance synthesized in one location, usually in small amounts, and then transported to another location within the organism where if affects growth and differentiation.

Humic acid: Mixture of dark-colored organic materials of indefinite composition extracted from soil with dilute alkali and precipitated by acidification.

Humification: Breakdown of organic residues to humus.

Humus: Fraction of soil organic matter remaining after most plant and animal residues have decomposed; usually dark colored. The chemical composition of humus is very different from that of the original parent compound. Humic substances (1) help the soil retain water, (2) increase the cation exchange capacity, and (3) stabilize soil pH.

Hydraulic conductivity (K): Flow-rate constant, expressed in centimeters/second, which indicates the ability of soils to transmit flowing water. Values of K commonly range from 1×10^{-1} cm/sec in sands to 1×10^{-9} cm/sec in tight clays.

Hydroponics: Commercial production of plants in sand or gravel cultures. The sand or gravel is relatively inert, providing mechanical support for growth, and nutrients are supplied by liquid solutions.

Hypothesis testing: Statistical technique which uses experimentation to support or reject a (null) hypothesis, formulated before the experiment, with a certain level of confidence.

I

Incorporation: Mixing of a fertilizer or chemical into nursery soil before sowing.

Indeterminant growth: Shoot growth resulting from the successive periods of initiation and elongation of apical meristem cells during the growing season without extended periods of rest (compare with **sdeterminant growth**).

Inductive reasoning: Making inferences or drawing conclusions, based on a limited number of observations, about a wider sphere of interest (compare with **deductive reasoning**).

Infiltration rate: Rate at which water can be absorbed into a soil surface. Infiltration rate can be altered by nursery practices influencing the porosity and structure of surface soils. A soil with a poor infiltration rate is subject to surface runoff and erosion.

Inoculation: Process of introducing microorganisms for some beneficial effect, such as the addition of *Rhizobium* bacteria to legume seed or of mycorrhizal fungi to nursery seedbeds.

Inoculum: Portions of a pathogen (e.g., fungal spores) capable of causing infection or initiating mycorrhizae upon contact with the host.

Interception system (drainage): Type of closed drainage system located below the water-table level, intended to remove water before it enters the soil. The system may vary from a simple line of pipes positioned at the source of seepage to a complex grid pattern and is usually used in small land areas (compare with **relief system**).

Ions: Atoms or groups of atoms which are electrically charged, i.e., cations or anions.

L

Lateral pruning: Root-culturing technique in which blades or colters are passed between drill rows to sever long lateral roots. The purpose of lateral pruning is to facilitate lifting, stimulate root growth and fibrosity, and retard height growth (compare wifih **box pruning**).

Leaching: Downward movement of materials in the soil solution. Soluble nutrients such as nitrate are often leached out of the seedling root zone.

Lifting window: Time period of the year believed to be the best for harvesting seedlings from the seedbed, i.e., when seedlings are most resistant to handling stresses and when subsequent survival and growth potential upon outplanting are high. The lifting window will vary from year to year depending on seed source, nursery location, and cultural regimes used before lifting.

Lignification: Deposition of lignin (complex aromatic compounds) in the cell walls of sclerenchyma, xylem vessels, and tracheids, making them rigid.

Lime (calcium) requirement: Amount of agricultural limestone required per acre to raise the soil pH to an optimal value for seedling growth; usually calculated for a soil depth of 15 cm or per 910,000 kg of soil.

Liming: Addition of calcium, sometimes including magnesium (dolomite), in the form of calcium carbonate, ground limestone, or hydrated lime to furnish elements for plant growth and to neutralize soil acidity.

Loam: Textural class for a soil having moderate amounts of all three soil separates-sand, silt, and clay.

Luxury consumption: Absorption of excess amounts of nutrients beyond those necessary for optimal growth.

M

Macronutrient: Chemical element necessary for plant growth in large amounts (usually greater than 1 part/500 in the plant); often supplied artificially in fertilizers (compare with **micronutrient**).

Maintenance dressing: Fertilizer application which functions to keep soil macronutrient and micronutrient levels adequate during the. crop rotation.

Material attributes: Individual measurements of one aspect of seedling quality which can indicate physiological condition, e.g., leaf osmotic potential, root starch concentration, and shoot:root ratio. Performance attributes reflect the sum total of material attributes.

Matric potential: Largest component of total soil-water potential; caused by capillary action and attraction of water by soil particles. Matric potential can be measured with tensiometers and is usually expressed in negative pressure units such as - bars.

Measurement plot: That part of the experimental plot in which the observational units are measured.

Megagametophyte: Haploid generation portion of a seed plant representing the female contribution to the developing embryo.

Metabolism: Chemical processes comprising the synthesis and degradation of constituents within an organism.

Micronutrient: Chemical element necessary for plant growth in very small amounts (less than several parts per million in the plant). Micronutrient fertilizers are not normally needed except under soil conditions such as excessively high or low pH (compare with **macronutrient**).

Milliequivalent (meq): One milligram (mg) of hydrogen or the amount of any other ion that will combine with it. Milliequivalents are units used in cation exchange capacity and fertility calculations. For example, 1 meq of a calcium ion (Ca^{++}) is computed as its atomic weight in grams (40) divided by the valence (2), or 20 mg.

Mineralization: Conversion of organic elements to the inorganic state as a result of microbial decomposition.

Mineral soil: Soil consisting largely of mineral matter, with organic matter usually less than 20%.

Mineral spirits: Derivatives of naphthenic petroleum, containing 10 to 20% aromatic hydrocarbons, used as a solvent for some pesticides; sometimes used in tree nurseries as a contact herbicide.

Monocot (Monocotyledoneae): Smaller of the two classes of flowering plants (Angiospermae), distinguished from the larger class (Dicotyledoneae) by the presence of a single leaf (cotyledon) in the embryo and by other structural features, e.g., parallel veined leaves; includes grasses, lilies, orchids. Certain herbicides are effective against monocots but do not harm dicots.

Mulch: Layer of plant residues or other material (e.g., plastic film, paper fiber) spread upon the soil surface to protect soil, seeds, or plant roots from the effects of freezing, evaporation, crusting, etc.

Mycorrhiza(e): The biological association, usually symbiotic, between plant roots and particular fungi.

N

Narrow leaf species: Those plants classified as Monocotyledoneae; characterized by having narrow, parallelveined leaves (compare with **broadleaf species**).

Natural regeneration: Reforestation of a stand by natural seeding (compare with **artificial regeneration**).

Nitrification: Biological process in which (1) ammonium is oxidized to nitrites, and (2) nitrites are further oxidized to nitrates.

Nonselective pesticide: Material that is toxic to a wide range of pests or to more than one plant or animal.

Null hypothesis: Specific hypothesis about a population that is being investigated by analyzing data from a sample of that population. For both statistical and biological reasons, elements under investigation are hypothesized to have *no* effect on the response variable thus, the use of "null."

O

Observational unit: Item to be measured within an experimental plot (e.g., tree seedling).

Operational trial: Preliminary inquiry in which each treatment is applied to only one plot (i.e., treatments are unreplicated) (compare with **designed experiment**).

Organic matter: The complex interaction of (1) plant, animal, and microbial residues in various stages of decay, (2) humus, and (3) live organisms. Organic matter increases the buffer capacity, cation exchange capacity, and water retention of the soil and provides a substrate for microbial activity.

Organic soil: Soil usually containing 20% or more organic, matter.

Ornamentals: Plants, including trees, shrubs, and flowers, which function to beautify homes, gardens, and lawns; refers to stock used for landscaping rather than wild land plantings.

Osmotic potential: Pressure that would have to be applied to a solution to prevent water from moving from the less concentrated solution to the more concentrated one when two solutions are separated by a semi-permeable membrane. Soil osmotic potential refers to solute concentration in the soil solution; plant osmotic potential refers to solute concentration of the cell sap.

Outplanting: Planting of seedlings on a forest site.

Oxidation: Chemical process of combining with oxygen; removal of hydrogen or electrons.

P

Parasite: Organism that lives on other living organisms.

Pathogen: Specific agent (usually fungus, bacterium, virus, or nematode) that can cause infectious disease.

Peat: Largely undecomposed or slightly decomposed organic matter accumulated under conditions of excessive moisture and low oxygen availability; soil amendment used to increase soil organic matter and lower soil pH.

Perched water table: Surface of a local zone of water saturation held above the main body of ground water by an impermeable layer, usually clay or rock, and separated from the main body of ground water by an unsaturated zone.

Percolation rate: Downward movement of water through the soil, particularly the down ward water flow in saturated or nearly saturated soil. Percolation rate is used to calculate the internal drainage requirements of a soil.

Perennial: Plant that continues growing from year to year. Tops may die back in winter but roots or rhizomes persist (compare with **annual, biennial**).

Performance attributes: Attributes of seedling quality measured by assessing the performance of seedlings subjected to environmentally controlled test conditions, e.g., root growth potential and frost hardiness. Performance attributes reflect the sum total of material attributes.

Permanent wilting percentage: Water content of a soil when indicator plants growing in that soil wilt and fail to recover when placed in a humid chamber; varies with plant species but is generally considered to occur at approximately – 15 bars water potential.

Permeability (soil): Soil attribute that enables water or air to move through it; determined by soil porosity.

pH: Numerical measure (negative logarithm of the hydrogen ion activity) of the acidity or alkalinity in a soil or solution. A pH reading of 7 is neutral for soils measured in water paste.

Phenotype: Morphological appearance of an organism. The phenotype results from the interaction of genotype and environment (compare with **genotype**).

Photoperiodism: Plant response to relative length of day and night. Temperature and photoperiod are the principal environmental factors affecting plant dormancy.

Photosynthesis: Production by plants containing chlorophyll of organic compounds from water and carbon dioxide, using energy absorbed by the chlorophyll from light.

Phytotoxic: Causing injury or death to plants.

Plant moisture stress (PMS): Measure of plant water status; equal to the absolute value of plant water potential. PMS is an integrated index of the current moisture status of a plant, and is influenced by soil moisture status and evaporative demand of the atmosphere.

Plant water potential: Current water status of a plant; consists of two components, turgor potential and osmotic potential, and is measured in negative pressure units (– bars).

Plasmolysis: Shrinkage of cell protoplasm away from the cellulose wall due to osmotic withdrawal of water.

Plug plus one (p+1): Transplanted seedling that was started in a container and then transplanted to the field, usually for 1 year (compare with **stock type**).

Point estimate: Number that estimates a certain quantity in the population of interest (e.g., treatment mean, standard deviation).

Pore space: Total space not occupied by soil particles in a bulk volume of soil.

Porosity (soil): Volume of total soil bulk not occupied by solid particles, expressed as a percentage. Percent porosity equals the volume of pores divided by total soil volume.

Postemergence: Time period *after* crop plants or weeds emerge through the soil surface.

Preemergence: Time period *before* crop plants or weeds emerge through the soil surface.

Preplanting treatment: Application of, e.g., herbicide or fertilizer before a crop is planted.

Profile (soil): Vertical section of soil extending through all of its horizons and into the parent material.

Prolepsis shoots: Shoot growth resulting from the expansion of lateral buds at the base of the terminal bud or on lateral shoots. Prolepsis shoots may be more susceptible to winter injury because of inadequate time to harden off.

Propagule: Any part of a plant that is capable of growing into a new organism.

Q

Quiescence (seedling): Period of physiological inactivity preceding true dormancy when plants still can grow if environmental conditions are suitable.

R

Radiation frost: Frost occurring when large amounts of heat in soil and plants are dissipated into the atmosphere, allowing temperatures near the ground to reach the freezing point. This condition occurs on cloudless nights with little or no wind (compare with **advection frost, wind frost**).

Radicle: Root of the embryo in seed plants.

Randomization: Assignment of treatments to a set of plots such that all plots are equally likely to receive any treatment.

Relief system (drainage): Type of closed drainage system located below the water-table level, designed to drain already saturated soils. The system may consist of a simple pipeline or a complex interconnecting network (compare with **interception system**).

Replication: Repetition of a treatment in an experiment.

Residual sodium carbonate (RSC): Irrigation-water quality index that measures the difference between the sum of calcium and magnesium ions and the sum of carbonate and bicarbonate ions. RSC values are used to determine the effect of salts on soil permeability (compare with **adjusted sodium adsorption ratio, ASAR**).

Respiration: Metabolic process of taking oxygen from the environment to produce energy and release carbon dioxide (in organisms); oxidative breakdown of fuel molecules and subsequent release of energy (in cells).

Response variables: Characteristics which an experiment and its treatments are designed to test (e.g., height, caliper).

Ripping: Cultural practice used to ameliorate compacted subsoils by pulling shanks through the soil at a depth of 40 to 80 cm. Usually, the shanks are then pulled at right angles to the first pass to produce a grid pattern (compare with **subsoiling, chiseling).**

Rolling: Cultural practice used before sowing to ensure good contact between seeds and soil particles. A cylindrical roller is passed over the land to firm the soil without causing a great deal of compaction.

Root culturing: General term for those nursery cultural practices designed to modify seedling root growth (e.g., undercutting, wrenching).

Root-growth capacity (root-growth potential): Physiological capability of a plant for producing new roots under optimal environmental conditions for root growth. High root-growth capacity is thought to be one of the best indicators of seedling field survival.

S

Salt: Water-soluble chemical compound that either is found naturally in the soil or is applied as fertilizer. In soil solutions, salts are generally dissociated into cations (e.g., calcium, sodium, potassium) and anions (e.g., sulfates, nitrates, bicarbonates). High salt levels are detrimental to plant growth. Conifer seedlings are especially sensitive to salts.

Sand: Soil particle between 0.05 and 2.00 mm in diameter; soil textural class characterized by a predominance of sand particles.

Saprophyte: Organism that lives on dead or decaying organic matter.

Scarification (land): Type of site preparation in which the duff and litter layers of the forest floor are removed or the mineral soil is mechanically mixed with the organic surface layer.

Scarification (seed): Process of scratching the seed coat with abrasive material to improve germination of seeds with hard seed coats which are relatively impervious to water.

Seedbed: Elongated strip of prepared soil in which seeds are sown and seedlings raised.

Seedbed density: Number of seedlings growing in a seedbed, expressed relative to area (e.g., number per square meter or foot) or lineal measure (e.g., number per lineal meter or foot).

Seedling: Young tree propagated from seed.

Seedling quality: Potential of a seedling to survive and grow successfully after outplanting.

Seedlot: Quantity of seeds from a particular location and elevation (seed zone) which are reasonably similar or uniform in quality. The identity and integrity of each seedlot (one of the basic divisions in seedling recordkeeping) are maintained during seed storage and during the nursery production period.

Seed protectant: Pesticide applied to seed before planting to protect seeds and new seedlings from diseases and insects.

Seed purity percent: Proportion of the total seedlot that is seed and not debris, expressed as a percentage of the seedlot weight.

Seed zone: Area of similar environmental conditions. Plants originating from the same seed zone are believed to be similarly adapted to the environment.

Serotinous: Cones which remain closed after maturing and which do not release seeds until several years after reaching maturity or exposure to high temperatures (e.g., some ecotypes of lodge pole pine).

Shale: Flat, layered rock consisting of consolidated clay and silt.

Shippable percent: Percentage of seedlings remaining at the end of the nursery growing period which meet certain size and form specifications (compare with **tree percent, yield percent**).

Silt: Soil particle between 0.05 and 0.002 mm in diameter; soil textural class characterized by a predominance of silt particles.

Sludge: General term for solid wastes, usually collected by sedimentation from water. Sludge is derived from many sources including agricultural wastes, brewery and cannery wastes, and sewage.

Soil horizon: Layer of soil approximately parallel to the land surface and differing from adjacent genetically related layers in physical, chemical, and biological properties or characteristics such as color, structure, texture, consistency, amount of organic matter, and degree of acidity or alkalinity.

Soil-moisture retention curve (soil-water characteristic curve): Curve depicting the relationship between soil matric potential and soil water content; varies with soil textural class and is usually calculated by soil-testing laboratories. This curve aids nursery managers in regulating soil matric potential and therefore irrigation scheduling.

Soil test: Chemical or physical analysis of a soil to determine texture, acidity, total salt concentration, or concentration of nutrient elements.

Soil water potential: Amount of work that a plant must exert to absorb water from the soil, usually expressed in units of negative pressure (- bars). It is composed primarily of matric potential but can also have as components osmotic potential, pressure potential, and gravitational potential.

Sowing: Process of placing seeds in the seedbed at a specific depth and density.

Specific gravity: Ratio of weight of a volume of matter to weight of an equal volume of water at a specific temperature.

Standard treatment: Treatment which simulates the operational procedures of a current practice (compare with **control treatment).**

Statistical Inference: Process of making generalizations, based on sample data, about a population or wider sphere of interest.

Statistics: Use of mathematical techniques to draw conclusions from a limited number of observations.

Stock type: Seedling classification, usually by age and location in the nursery-e.g., 1+0, 2+0, etc. The first of the two digits represents the number of growing seasons spent in the seedbed, the second digit the number of growing seasons spent in a transplant bed. " 1+0 for 1+1 " or "2+0 for 2+1 " means seedlings grown for transplanting, often under specific cultural conditions (e.g., high seedbed density).

Stratification (seed): Treatment applied before germination to overcome seed dormancy. Cold stratification consists of placing seeds in an environment of cold temperature, sufficient moisture, and oxygen for a specified time period.

Subsoiling: Tillage of subsurface soil without inverting it, to break up dense soil layers that restrict water movement and root penetration.

Summer annuals: Plants which germinate in the spring, do most of their growing in the summer, produce flowers or seeds, and then die in the fall of the same year (compare with **winter annuals).**

Surfactant: Chemical agents (e.g., spreaders, detergents, wetting agents) added to pesticides to make mixing easier and to assist application of a solution and adherence to the treated surface.

Symbiosis: Association of two dissimilar organisms, usually referring to cases in which the relationship is beneficial to one or both organisms.

Systemic: Entering and then acting within an entire organism; used especially to describe the action of pesticides or diseases within a plant.

Systemic herbicide: Herbicide which is absorbed by and then distributed within a plant, as opposed to one which func functions only on contact with the plant's surface (compare with **contact herbicide).**

T

Table pruning: Pruning of seedling roots at the time of grading and packing.

Tensiometer: Instrument for measuring the matric potential of soil water, often used for monitoring irrigation.

Tilth: Physical condition of soil as related to its ease of tillage, fitness as a seedbed, and impedance to seedling emergence and root growth.

Tissue analysis: Chemical analysis of plant tissue to determine the concentrations of essential elements in the plant.

Top dressing: Application of chemical or fertilizer after a crop has been established.

Top pruning: Clipping of seedling terminal leaders with a sharp blade to alter shoot:root ratio, facilitate handling, achieve uniformity in crop size, and control height growth.

Transpiration: Process of water movement through a plant to the atmosphere as a result of evaporation of water from leaves.

Transplant: Cultural practice of moving seedlings from one bed to another to promote additional growth. Also, a seedling after it has been lifted and then replanted one or more times in the nursery.

Transplant shock: Reduced growth rate of a young tree after it has been transplanted or outplanted.

Tree percent: Number of seedlings, irrespective of size or form, in a nursery bed at lifting compared to the number of viable seed sown, expressed as a percentage (compare with **yield percent, shippable percent**).

Turgor potential: One of the main components of plant water potential; reflects a positive force exerted inward by the cell wall.

U

Undercutting: Root pruning in the nursery bed using a sharp blade drawn parallel to the soil surface at a regulated depth to stimulate root growth and fibrosity.

V

Vesicular-arbuscular (VA) mycorrhiza(e): Group of mycorrhizae in which the fungal hyphae form two characteristic structures—vesicles and arbuscules—which are both intercellular and intracellular. Vesicles are saclike storage and reproductive structures, whereas finely branched arbuscules facilitate nutrient exchange. Though not visible to the naked eye, VA mycorrhizae can be seen by staining infected cells and then inspecting by microscope.

Viability: Ability of a seed to germinate and grow under a given set of conditions; usually estimated by germination percent or other tests.

W

Water content (soil): Index of soil moisture status, calculated as the amount of water lost from the soil upon drying to constant weight at 105°C; usually expressed as the weight of water per unit weight of dry soil.

Waterlogged: Saturated with water. Waterlogged soil, which may result from a high water table caused by over irrigation, seepage, or inadequate drainage, is detrimental to plant growth.

Water table: Upper surface of the ground-water level, below which the soil is saturated with water.

Wettable powder (WP): Powder formulation of a pesticide which contains a wetting agent so that it will readily form a suspension in water.

Wetting agent: Compound added to a pesticide solution causing the spray droplets to spread and more thoroughly wet the leaf surface.

Wind frost: Frost occurring when winds in excess of 4 mph from cold regions displace warmer air. Wind frosts can occur day or night and are not necessarily dependent on topography.

Winter annuals: Plants which germinate in the fall or early winter, exist in a rosette form over the winter, do most of their growing the following spring, and then die after flower-ing and producing seeds.

Winter burn: Type of cold injury to foliage. Foliage is warmed above freezing by the winter sun during the day (even though air temperature is below freezing), then refreezes after sunset.

Winter desiccation: Type of foliage injury which occurs on warm days when the ground is frozen: actually a type of physiological drought caused by excessive transpiration when frozen soils prohibit water absorption.

Winter scald: Type of cold injury to tree bark. Bark is warmed above freezing by the winter sun during the day (even though air temperature is below freezing), then refreezes after sunset.

Wrenching: Passing of an *angled* horizontal blade beneath the soil surface of the nursery bed at a specified depth to cut newly penetrating roots and to loosen and aerate soil. Wrenching is used to stimulate root growth and fibrosity and to regulate seedling growth.

Y

Yield percent: Number of trees which meet a specific size criterion, regardless of form; expressed as a percentage. These seedlings may have multiple tops or damage from insects, disease, or other agents—characteristics that may make them unacceptable for shipping.